Praise for

THE ASTOR O

"Novelistic. . . . Vividly gothic. . . . It's a trick to tell a story this rich and complicated through the eyes of a child without losing the subtleties of character and nuances of history, but Aldrich pulls it off with aplomb." —*Kirkus Reviews* (starred review)

"Aldrich bears witness to the tail-end of the disintegration of that most storied of American dynasties, the Astors. . . . This unflinching memoir of childhood chaos and neglect is relieved and enlivened by Aldrich's wittily sharp observations and her obvious affection for her peculiar relations." —*Booklist*

"The book is a meditation on a way of life . . . a bit like getting lost in a world somewhere between fantasy and nightmare, where the ghosts of a particular type of antique American greatness confront the realities of the modern world." —Smithsonian

"What's most notable here is the melancholy that Aldrich elicits throughout, a sensitivity towards time and fate reminiscent of a painted landscape in which a shepherd stops to rest his flock among the forgotten, grown-over ruins of a once mighty castle." —Daily Beast

"A poignant story that lays bare a woman's search for self-explanation." —*Weekly Standard*

"Alexandra Aldrich spins gold in *The Astor Orphan*."

—*Buffalo News*

"*The Astor Orphan* is . . . gorgeous, ugly, sad, bohemian, and only mildly sordid or scandalous by TV or literary standards."

—*Star-Tribune* (Minneapolis)

"Throughout *The Astor Orphan,* you will laugh, you will cry . . . but mostly you will come to appreciate the characters that colored Aldrich's life. . . . In this striking memoir, we learn how to succeed in face of failure."

—digboston.com

"A beautifully-rendered family saga—full of fires, affairs, aristocrats, and illegitimate children. At the center is an endearing heroine, whose eccentric childhood on the *Grey Gardens*–style Rokeby estate would make Dickens gasp. It is a story of lost grandeur, familial love, and the determination of one brave young woman to make a better life for herself. A splendid memoir."

—Jennifer Vanderbes, author of *Easter Island* and *Strangers at the Feast*

"Alexandra Aldrich is a member of the storied Astor clan, but her 1980s childhood on the family's decaying Hudson Valley estate was a far cry from that of her aristocratic predecessors. In this evocative debut memoir, Aldrich astutely portrays a colorful cast of aunts, uncles, cousins and hangers-on—clinging to the family legacy long after the money is gone. One can't help but cheer as she breaks away from the others to make a name for herself."

—Elliott Holt, author of *You Are One of Them*

THE
ASTOR
ORPHAN

A MEMOIR

Alexandra Aldrich

ecco

An Imprint of HarperCollinsPublishers

Grateful acknowledgment is made for use of the following photos:

pp. 5, 31, 93, 167, 191: Ralph Gabriner; pp. x-xi, 7, 117, 119, 127: courtesy of the Margaret Livingston Aldrich Papers; p. 11: courtesy of *Town & Country* magazine; pp. 19, 47, 59, 177, 197, 205, 215, 229, 241, 247, 257: Ania Aldrich; pp. 41, 71, 147, 227: Charles Tanguy; pp. 57, 139: © by China Jorrin; p. 81: The City College of New York, CUNY; p. 95: Georgiana Warner; p. 109: courtesy of Michele Michahelles; p. 135: courtesy of the author; pp. 157, 179: Sarah Stitham.

This is a work of nonfiction. Though all the characters' names have been changed, the events and experiences detailed herein are all true and have been faithfully rendered as I have remembered them. Though one-armed Roy is a composite of two family friends, all other characters are faithfully rendered portraits of real people. Time lines have been compressed to retain narrative flow. I rendered the dialogue as accurately as I could, but as none of it was electronically recorded, I cannot guarantee that it is a word-for-word representation of conversations that took place more than twenty years ago.

HarperCollins books may be purchased for educational, business, or sales promotional use. For information please e-mail the Special Markets Department at SPsales@harpercollins.com.

A hardcover edition of this book was published in 2013 by Ecco, an imprint of HarperCollins Publishers.

FIRST ECCO PAPERBACK EDITION PUBLISHED 2014.

Designed by Suet Yee Chong

Library of Congress Cataloging-in-Publication Data has been applied for.

ISBN 978-0-06-220795-1

14 15 16 17 18   OV/RRD   10 9 8 7 6 5 4 3

To my parents,
Rokeby's current guardians

CONTENTS

PART FOUR

All in a Summer's Plunder

PART FIVE

Other Exiles

PART SIX

In Search of Self

ILLUSTRATIONS

THE · AMERICAN · ANCESTRY · OF · JO...

Compiled for their Descenda...

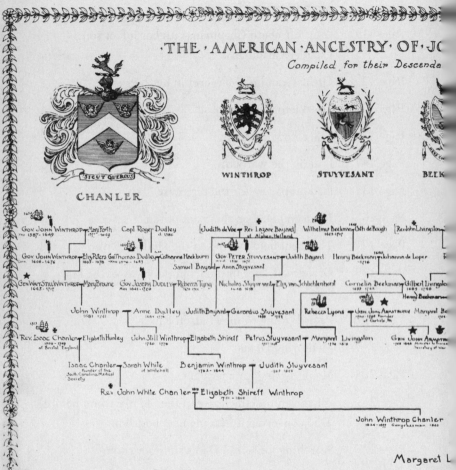

WINTHROP STUYVESANT BEEK...

CHANLER

Margaret L...

·NTHROP·CHANLER·AND·MARGARET·ASTOR·WARD·

rence Grant White – Christmas MCMXXI·

LIVINGSTON   **SCHUYLER**   **GREENE**   **WARD**

Brandt van Slichtenhorst
Director of Renesselaerwyck.

John Greene ⊤ Joan Tattersall
1597–1659

ROGER WILLIAMS ⊤ Mary Warnard
1599–1683   –1676

terse Schuyler ⊤ Margaret van Schlichtenhorst   Eliz. Barton ⊤ Thos Greene   Maj. John Greene ⊤ Anne Almy   John Sayles ⊤ Mary Williams   John Ward
1628–1683   1628–1711   1620–1708   1627–1709   1633–1681   1613–1681   1619–1698

Alida Schuyler   Susanna Holden ⊤ Benj. Greene   Mary Gorton ⊤ Samuel Greene   Job Greene ⊤ Phebe Sayles   Isabel Sayles ⊤ John Tillinghast   Amey Billings ⊤ Thos Ward
1656–1709   1670–1704   1678–1732   1656–1748   –1744   1678   1615–1000   1641–1689

Margaret Howarden ⊤ PHILIP LIVINGSTON   Benj. Marion ⊤ Judith   Cath. Greene   Gov. Wm. Greene   Deborah Greene   Simon Ray   Mary Tillinghast   Gov. Rich. WARD
                   Signer of The   1690–1771   R.I. 1690–1748   1640–1708   1672–1708   1689–1767   R.I. 1689–1763
                   Declaration of Independence

R. Livingston   Gabriel Marion ⊤ Esther Cordés   Gov. Wm. GREENE ⊤ Catherine Ray   Ann Ray   Gov. SAMUEL WARD
                                A Relative of   R.I. 1731–1809   1717–1760   1725–1774
                                Charlotte Corday

gston   John Jacob Astor ⊤ Sarah Todd   GEN FRANCIS MARION ⊤ Esther Marion ⊤ Benj. Cutler   Phoebe Greene   Lt. Col. Samuel Ward
         1763–1848   –1834   1735–1795   1760–1828   1756–1832

trong ⊤ William Backus Astor   Julia Rush Cutler ⊤ Samuel Ward
        1792–1875   1790–1824   1786–1839

Emily Astor ⊤ Samuel Ward
1819–1841   1814–

ret Astor Ward
1838–1870

Chanler ⊤ Richard Aldrich.

PROLOGUE

"The Declaration of Independence was drafted by—among others—our ancestor Robert Livingston," I lectured my younger cousin Maggie as we crunched through the snow on our way to the pond. It was mid-December.

"You forgot about the tea." Maggie yawned.

"I didn't forget. That isn't what I'm discussing. But I can tell you about the tea if you'd like."

"I already know."

I was ten and used instruction of various kinds to assert my authority over my younger cousins—Maggie, six, and Diana, five—as frequently as I could. I was determined that they shouldn't grow up to treat me with the same disdain with which the family treated my father. But it seemed the damage had been done.

Maggie eyed me slyly. "Who bought you that new down vest?"

"Grandma did."

"Grandma says your daddy can't get a job."

"That's not true. He fixes things, plows and mows the fields . . ."

"He's poor. That's why Grandma buys you clothes."

I had no answer.

Our skates, tied together so we could carry them on our shoulders, went *clickety-clack* as they swayed with the rhythm of our walking. We used secondhand figure skates that Grandma Claire picked up at church rummage sales; they were, therefore, only our approximate sizes. *Too big won't matter, because you wear layers of socks, anyway. And you'll grow into them soon enough.*

Mom was with us, her red two-piece snowsuit swishing as she strode along. Mom came to life in the dead of winter. Her Polish cheeks grew rosy, and her breath grew strong with the exertion of hiking through snow or skiing cross-country. She seemed to be back in her native element, the wintry Carpathian paradise where she'd grown up.

Mom skated onto the ice first. With her snow shovel, she rode along the surface, pushing aside the light, fluffy snow. She started a counterclockwise path from the center, creating an ever-widening spiral.

We three hobbled, trying to get our balance. Diana could only take halting steps, couldn't yet push and glide. Maggie was a bit more adept: she pushed and glided, lost momentum, then pushed again, gliding a few feet before she stopped.

The ice was bumpy and mostly black, with some white

swirls where air had been caught, so we seemed to be floating above the Milky Way.

Suddenly, Diana started hollering and pointing at the ice in front of her.

"It's the devil! Ahhhhh!"

Accustomed as we were to Diana's tantrums, none of us felt any alarm. But when I skated over to her, I saw two faces frozen in the ice.

Dad's goats were staring up at us, with expressions of what was either panicked terror or ecstatic delight. Their toothy, open mouths could have been calling out either "Save us from the black abyss!" or "Ah, wow, the universe is so vast and miraculous!" The rectangular irises of their eyes were popped wide, and their front hooves rose as if they were reaching for a hand up.

"So *that's* where they went," said Mom stoically. "Don't tell your father."

"I want to go home!" Diana whimpered.

None of us wanted to continue skating; it didn't seem right to skate over the faces of these creatures that seemed still to be alive, their eyes and mouths wide open.

So we walked back up the hill to Grandma Claire's house. The crisp snow held our skates in balance, and when we came to the driveway, we had to bend our ankles and walk on the outer edges of our feet.

Mom helped Maggie and Diana unlace their skates on Grandma Claire's front stoop.

"Back so soon?" Grandma Claire poked her head out the storm door.

"There are devils in the ice," Diana informed Grandma Claire.

Grandma laughed. "Devils, really?"

"She means the dead goats, Grandma. They're in the ice."

"Oh! Ugh! Oh, dear me." Grandma Claire grimaced. "That sounds just frightful!"

Even more frightful was that I knew Grandma Claire had killed them.

PART I

ROKEBY DREAMS

CHAPTER ONE

THE LEGACY

As one tops the final rise of the gravel carriage drive, the eastern wall of the forty-three-room mansion appears. Even from a distance, one can see the brown water stains that streak the mansion's off-white stucco walls and the missing slats in the peeling black shutters that edge its long-paned windows. Layers of slate tiles cover the mansard roof like the scales of an armadillo.

The house's wide stone steps lead up to the porch and the faded frescoes that adorn its sandstone walls. One fresco depicts the Algonquin who dwelled on this land before it was granted to the Scotsman Robert Livingston Sr. by King James II in the 1680s. Another portrays Napoléon giving a flock of merino sheep to General John Armstrong Jr.—U.S. minister to France from 1804 to 1810, secretary of war in Madison's cabinet, and the man who built this house.

Known to all as the "big house," this mansion last rattled with life when the eleven Astor orphans, the great-grandchildren of William Backhouse Astor Sr., roamed free here, wild, willful, and beyond their guardians' control. The big house, with its high ceilings, vast open spaces, and secret niches so tempting to children, has never been conducive to discipline. The 450 acres of land surrounding the house have always served as a buffer against the outside world.

It was William Backhouse Astor Sr.—son of John Jacob Astor and the richest man in mid-nineteenth-century America—who brought great wealth to Rokeby, when he married the daughter of John Armstrong Jr. and Alida Livingston. And it was Astor's granddaughter Maddie Astor Ward and her husband, John Winthrop Chanler from South Carolina, who orphaned the large brood when they both died of pneumonia in the space of two years.

These free-spirited Astor orphans left us, their descendants, our legacy: the house, its history and contents, and a sense of entitlement and superiority. They were the original eccentrics of the family, each one unconventional and adventurous.

Of the eleven orphans, eight lived into adulthood. Lewis Stuyvesant Chanler—"Uncle Lewis"—reportedly saved ten men from the electric chair as one of New York's first pro bono attorneys, and later served as lieutenant governor of New York. William Astor Chanler—"Uncle Willie"—was Teddy Roosevelt's classmate at Harvard and fellow explorer, and reportedly hobnobbed with Jesse James in the Wild West, worked as a gunrunner in the Spanish-American War, and discovered sixteen new species of deer during his hunting expeditions in East Africa. Robert Chanler—"Uncle Bob"—was a well-known painter of murals and screens, and onetime sheriff of our town. John Armstrong Chanler— "Uncle Archie"—was incarcerated in a mental hospital by his brothers for his bizarre behavior. Winthrop Chanler— "Uncle Wintie"—was also a hunter, and his daughter Laura married the son of Stanford White. Aunt Alida Emmet was presented to the queen of England as a girl and had the distinction of being the last living member of Caroline Astor's original list of the four hundred most elite members of New York's high society. Aunt Elizabeth Chanler married the great American essayist John Jay Chapman, and was the subject of a portrait by John Singer Sargent that is kept at the National Gallery in Washington, DC. Finally, my greatgrandma Margaret Chanler became the sole owner of Rokeby as a young woman, after buying out her siblings.

Evidence of the orphans' exploits surrounds us and offers us a standard to live by.

Today the Astor money, which has supported generations of aristocrats ill equipped to earn or invest, is gone. While that undisciplined, orphaned spirit still abounds, it is depressed by the house's sterile air of preservation. Aside from its continuous deterioration, the house has changed little since Great-Grandma Margaret's death in 1963.

To keep the house as it was then, we sacrifice any resources that might have been invested in current generations. In return, the house gives each of us—the impoverished descendants—an identity. And we live off the remains of our ancestral grandeur.

CHAPTER TWO

GUARDS OF ORDER

TOWN & COUNTRY
NEW SERIES OF THE HOME JOURNAL

Volume 57, Number 43   New York, Saturday, January 3, 1903   Price Ten Cents

MISS MARGARET LIVINGSTON CHANLER

Miss Chanler has done much more than the usual young woman whose family and personal charm give her the right to claim every pleasure of the social world. She went with the Red Cross Society to St. Petersburg; has been associated with Clara Barton in her work; will be remembered for her philanthropic work in Pekin, and started the hospitals for soldiers in the Philippines. She is a sister of Mr. William Astor Chanler.

Taffeta rustled and crackled as my little cousin Diana ran, dragging Great-Grandma Margaret's black gown across the drawing room's radiant orange Oriental rug. My cousins and I had slipped from the mansion's servants' quarters into its formal front rooms—usually reserved for tours and parties—in order to practice for a play.

Beethoven rippled from my violin and over the room: lamps shaped like elaborate candelabras, folds of torn wallpaper, peeling paint, tattered lampshades, and warmly lit mirrors that towered over the room's two gray marble mantelpieces.

As the provider of background music, I was not dressed in a gown.

This, the drawing room, was the largest room in the house, a continent unto itself, with two Steinway grands— black and brown, one for each of my musical great-grandparents. The pianos slept, cheek to jowl, at one end of the room. At the other stood a gilt table with a bare-breasted cherub astride each of its front legs. The rest of the room was uncluttered, waiting for an audience to convene for a recital on the several armchairs upholstered in lime-green silk.

Diana was carrying some dolls from my extensive collection. Her short, dirty-blond hair had been chopped unevenly in a self-inflicted haircut. In one arm, she held my Jenny doll—a two-foot-tall, soft-torsoed toddler whose beige plastic folds of chubby flesh collected at her knees and elbows and whose eyelids rolled open and closed with a little clicking sound as she turned. In her other arm, Diana held both Eva, a wooden Polish doll with an itchy red woolen skirt and two ruddy brown braids, and the Russian soldier doll with a coarse olive army uniform and crumbling plaster face.

Maggie lay on a sofa with a torn floral slipcover, dressed in a gown of robin's-egg blue and inhabiting the role of our great-great-grandmother Maddie Chanler. Her eyes were closed and her arm was draped dramatically across her forehead.

"Come see your Mama Maddie now," Diana, in the role of Cousin Mary, said to the babies in her arms. She then placed them gently around Maggie.

"Thank you, Cousin Mary," Maggie said to her sister. Then she turned to her babies. "Your mama is dying. She has ammonia."

"It's *pneu*-monia," I called out over the music. We all knew the story from Maggie and Diana's father, who lectured historical societies as they toured the house.

"You see? I'm dying, and she's still correcting me!" Maggie momentarily lost her composure, then resettled into character. "Rokeby has always taken care of its orphans. My mother . . . What was my mother's name again?"

"Emily Astor," I called out over the decrescendo as a chip of ceiling paint fell onto the piano.

"My own mother, Emily Astor, died when I was just two years old. And my grandparents . . ." She paused.

"The William B. Astors," I called over the trill.

" . . . raised me here at Rokeby as their own daughter."

Diana bit her lip as she fished for the next line, then blurted out, "And they banished that vile father of yours!" Diana was playing Mary Marshall, a cousin who became the orphans' primary guardian. "Sam Ward! His name should not even be spoken in this house!"

"But what did my daddy do?" asked a languishing Maggie.

"He was a bohemian spendthrift!"

Hanging like an idol on our drawing room wall, a youthful Great-Grandma Margaret—one of the youngest of these bereft babies—looked on from her gilt frame. She was seated on the front porch in a raven-black gown, poised and self-righteous with her back ramrod straight and her thin, elegant hands resting on her full, bombazine-coated lap. It was the same gown Diana now wore.

GREAT-GRANDMA MARGARET WAS Rokeby's ancestral guard of order. In reaction to her undisciplined and tragic Rokeby childhood, she had developed rigid rules and unyielding opinions. As an adult, she never varied from her schedule of reading, meals, and visiting hours at teatime. These had been the pillars of sense and sanity for her, if not for her relatives, who had simply kept their distance—some by choice and others by force.

As the sole owner of Rokeby, Great-Grandma Margaret had had the power to banish any disorderly elements, namely family members who defied her standards and expectations. Her fanatically strict principles superseded any emotional ties with even the closest of family members. For her, the greatest threats to the family's respectability were divorce and religious conversion.

Among the banished were her favorite brother, Lewis (divorced), and her own daughter, Maddie (also divorced), as well as her sister Alida, who had chosen to become Catholic.

IN MY MIND, I, too, was a guard of order, perpetuating the family's image of class and refinement with my violin playing and outstanding academic record.

"Maggie, that dress should not be so tight fitting on you," I prompted as the girls gathered up the dolls after rehearsal. "You don't see anyone overweight in the portraits, do you?"

"I'm telling my mommy you called me fat!" She stamped her foot and stalked away. Maggie always went over my head to those with the real power.

I often wished we were orphans, with enough inherited money to live on. But money was the only thing we hadn't inherited.

I now picked up my violin and began to play again, this time loudly, just for the pleasure of hearing my own sound resonate through the massive room.

When I played my violin, there was no past, present, or future. The sense that we lived on the brink of disaster was suspended, as was the sickening feeling that there was nowhere else in the world we could possibly belong. When I played, all that existed for me was the firmness of my bow's horsetail hairs as they glided and bounced on the steel strings; my left hand gently cupping the violin's smooth wooden neck and rocking, sometimes slowly and other times intensely, with vibrato; my thumb sliding up and down the neck as my hand changed positions; the piney smell of rosin powder as it floated off the dancing bow hairs; the deep, low notes, so solid and strong, and the high notes, vulnerable and brave; my fingers curling over the fingerboard as their callused tips pressed down and released with each note. How I admired the agility and obedience of those happy fingers!

"Olivia!"

Uncle Harry's voice woke me from my trance. Uncle Harry was Maggie and Diana's father; Olivia was their mother. He stood now on the western lawn, just outside the French doors, dressed in his usual suit and tie. His straight hair was smooth, oiled and brushed over to the side, and he wore round, steel-rimmed glasses.

Behind Uncle Harry rose a pillar of smoke from the front field. This was Dad—Uncle Harry's older brother—on his tractor, doing the spring mowing. Playtime was over. I had to prevent Dad and Uncle Harry from intersecting.

Each brother had taken on a secondary identity to supplement his endangered aristocratic one. Uncle Harry had adjusted well to the present, taking on a job as a civil servant in Albany. Dad remained at Rokeby full-time, in the dual roles of landlord and unpaid handyman.

As different as they were, each in his own way identified strongly with his aristocratic roots, and they shared a passion for keeping Rokeby's glory alive.

I now rushed to put my violin back in its case, which I had left in the octagonal library. As the shutters were kept closed to protect the books from daylight, it was dark—an effect intensified by its *faux bois* walls and ceiling. The walls were lined with approximately four thousand deteriorating, leather-bound books, and from the ceiling dropped an ominous hook, initially meant for a large chandelier. Stagnant because it was never aired out, the library harbored winter's chill all year long.

Diana was there, and I handed her my violin case as I ran off to join Dad on the tractor fender. On my way through the hall, I eyed the cobwebs that clustered around the legs of tables and chairs, the clumps of dried mud and cat hair scattered on the parquet floor, and the filthy white steps. All called me to dust, sweep, and wash them.

But the pull to Dad was stronger.

CHAPTER THREE

THE MENTOR

I raced out the double front doors, across the dusty circular driveway, and down the grassy hill toward Dad.

"Dad!" I shouted. But he couldn't hear me over the roar of the tractor engine and mower blades.

I was always on the lookout for Dad. I would scurry after him like a desperate pet, taking three steps for every one of his. "Wait up, Dad!" I'd cry as I followed him on his daily rounds of the property—digging trenches and laying new water, waste, or electric lines; doing mechanical work on one of the farm's tractors; or nailing new shingles onto a barn roof.

"You gotta keep up," he'd say, without turning around or adjusting his pace.

Dad was always chasing his own father—"Pop" to him, "Grandpa Dickie" to me. In Rokeby's soil and barns, on the iceboats, along the electric and water lines, lay his early childhood memories of time spent with his father. Mowing, riding, fixing—these were pursuits they'd shared before the liquor had swept Grandpa Dickie into permanent oblivion, then premature death. Dad would wake up at five, and then, over breakfast, write lists of things that needed to be done on the property. He never seemed to tire of it.

When I finally caught up with the tractor, Dad slowed down for me. I hopped on, hoping to keep him from driving up to the big house and running into his brother. Such confrontations usually resulted in Dad's being severely scolded by Uncle Harry for some activity he had undertaken at Rokeby without consulting his co-owner.

"Do you have any other work to do?" I shouted over the roar of the tractor engine.

"Trench—lower barn," he shouted back.

Dad's wiry hair was matted. His face and hands were smeared with grease and his face powdered with dust. The name tag on his used, blue work shirt read HANK. His gray eyes looked kind and wise.

I loved riding on the tractor fender like this, losing track of time in the vast greenness, grass seeds flying around our heads and scratchy bits of hay getting into our noses and eyes. As we drove, I watched a family of turkeys wobble past the rusted Rokeby windmill.

THE ASTOR ORPHANS' fancy New York relatives—especially their great-aunt Caroline Astor, *the* Mrs. Astor, the famous socialite—had viewed their country cousins as unrefined and had made extensive efforts to transform them into urban sophisticates, fit for high society. But the city Astors could not take the country toughness out of Great-Grandma Margaret.

As a young woman—long before she married the esteemed music critic Richard Aldrich—Great-Grandma Margaret significantly expanded the farm at Rokeby. When, in 1900, she inherited money from her great-aunt Laura (Astor) Delano, she used it to add three large barns to the Rokeby barnyard and purchase approximately fifty cattle, establishing a large-scale dairy farm, which she ran until her death more than sixty years later.

Dad parked the tractor on the farm road at the edge of the barnyard and approached the yellow backhoe that he'd use to dig the trench by the lower barn. It stood next to a lonely fuel pump in the cluttered yard, its long hoe curled up behind it like a tail.

The barnyard—although a sad version of what it had

once been—was still the heart of Rokeby, and Dad its pulse.

In the southeast corner of the barnyard stood a redbrick coach house designed by Stanford White. In the center was a U-shaped complex of white barns whose sections were still named according to their former purpose—the horse barn, the ox barn, the iceboat shed. Now the barns' siding was missing, roof tiles regularly blew to the ground, and doors hung off their hinges.

Just past the main barn complex stood a yellow farmhouse that, while badly in need of paint, was surrounded by a well-trimmed lawn littered with prickly Chinese chestnuts. This was where Sonny Day, Rokeby's seventy-year-old groundskeeper, lived. Sonny, who had worked at Rokeby since 1915, was one of the last human vestiges of the old farm days. These days, his duties were limited to mowing the lawns on an ancient riding mower and picking up the mail at a one-room post office down by the railroad tracks.

Dad called today's Rokeby "the funny farm." Nowadays, he used the barnyard to fix or store various broken tractor engines and appliances. Junk of all descriptions littered the yard: tires, old vacuum cleaners, broken fans, furnaces, lawn furniture, discarded washing machines, car transmissions. These were either donations from Dad's friends or leftovers from Rokeby's rental houses—converted outbuildings like the old greenhouse, the old "gardener's cottage," the creamery, and the "milk house"—whose tenants helped pay the property taxes.

"The hoe's gonna need some diesel. Hey, Roy!" Dad's voice searched among furnaces and mowers. "You here?"

"Yeah, Teddy! Over here!" A one-armed man with a scruffy black beard and a woolen cap now appeared from below the iceboat shed. "Just workin' on the bulldozer." Roy lived out in our woods, in an old school bus, with his family.

"Can you give the hoe some diesel?"

One-armed Roy was one of the many people Dad took under his wing and whom he taught to fix things on no budget, or entertained with stories about his adventures behind the Iron Curtain in the sixties. Like so many of Dad's "mentees," Roy had become an expert iceboater under Dad's tutelage.

Roy put fuel into the backhoe as I followed Dad into the horse barn. The stalls that had once been home to equine life now overflowed with detritus. As Dad poked around for some tools, I began picking up the garbage that cluttered the entrance and setting it aside in a pile.

"Wait just a minute! What are you doing with that?" Dad stopped me.

"Just throwing away garbage."

"That stuff is useful. I'll go through it myself later." Dad was a wall that stopped anyone from creating order in his barns.

Among these objects Dad hoped to find the essence of a Rokeby now lost. It was his dream to revive this essence, to return to the way things were when his pop was still alive.

My dream was to clean everything, so that it might one day look presentable to outsiders.

"Hey, Alexandra! Why don't you go out there and try to start up the backhoe?"

"Me? But, Dad, I'm only ten!"

" 'S about time you learned to drive a backhoe. I did at your age."

I couldn't say no. I had to prove to Dad that I was tough—a true farm girl. So I climbed into the seat. Its yellow stuffing poked out of the tears in its black plastic upholstery.

"Check that she's in neutral, now," Dad commanded from the horse barn's cluttered doorway. He had no patience for stupidity, so I didn't ask him how I was supposed to check. I just sat there, slumped and waiting.

"Go ahead and take the stick in your hand, and see if it's loose," he said in feigned exasperation at what he saw as my girlish tentativeness.

And I felt sorry that I wasn't a boy. I knew how much Dad had wanted a son—someone who could approximate his intimate relationship with Rokeby's land, buildings, and infrastructure when he was gone. As much as he let me tag along, I knew he would never see me as a proper heir.

The best I could do was to act like a boy so that Dad would teach me to sail an iceboat, weld, dig trenches with the backhoe, and drive a tractor.

I tried to move the stick shift. "It's not loose," I reported.

"Step on the clutch, then move the stick into the middle, where it'll wiggle freely."

I did as Dad instructed, excited to learn.

"Now keep your left foot on the clutch as you turn the key."

The key was slightly bent and hard to turn in the ignition. When I finally turned it, nothing happened.

"Needs starter fluid." Dad began scavenging amid the debris. He rolled a tire onto its side and some water gushed out of it. He kicked an empty oil can. Finally he found the starter fluid under a ripped blue tarp and sprayed something onto the uncovered engine that burned my lungs. I turned the key. As the backhoe roared to life, I felt a rush.

Just as Dad was about to show me how to put the machine into reverse, Grandma Claire's lemon-colored Plymouth sailed into the barnyard and stopped abruptly.

Grandma Claire opened the door and heaved her fragile body out of the low driver's seat. In her younger years, she had stood six feet tall, but now her back was badly hunched over. Her face was gaunt, her eyes sunk deeply into her skull. Her frizzy, once-black hair was cut short and had long since turned salt and pepper. She stepped gingerly around the piles of metal in her red espadrilles.

One-armed Roy darted out of sight.

Grandma Claire lived down the hill, past the barnyard, in what had once been the chauffeur's garage. After Grandpa Dickie died in 1961, Grandma Claire had the garage converted to store her, instead of Great-Grandma Margaret's '39 Mercury. And, for us, her granddaughters, Grandma Claire was indeed the chauffeur's replacement. She drove us to music lessons, the public library, doctors' appointments, stores and parties.

Grandma's skinny arms hung loosely from her torso, which was crunched up beneath the curve of her hunchback. She lifted them and waved at Dad, her son.

"Turn that thing off!" Her voice cracked with the desperation of a mother unable to control her child's errant behavior. "Who allows their child to drive heavy farm machinery?" Most of her *r*'s disappeared into her old New England accent. Her father's family had been New Englanders, and Grandma had spent her first ten years on Boston's North Shore.

I shut off the engine. Since Grandma Claire paid all the bills, I felt obligated to listen to her.

"Teddy, I demand to know what's going on here!"

"We were just testing out the backhoe, in preparation for some possible digging." Dad generally tried to disarm the fierce females in his life with meek acquiescence, though he was rarely successful.

The blood rose up Grandma Claire's neck like a rapid red tide. Her lips tightened, and she bared her large, once-glamorous teeth like a mad dog. "I suppose you think you're doing useful work here, do you? I suppose that digging up the place makes you feel better about not having a real job. . . ."

In true aristocratic fashion, Dad had never learned a profession. He had attended elite private schools, then Harvard and Johns Hopkins. After college, he traveled around Eastern Europe for six years, picking up five languages along the way. He had a gentleman's education, charm, and endless stories, but little professionalism.

The family viewed Dad's inability to conform to more middle-class norms as deliberate defiance, the mark of true failure, and the cause of all Rokeby's current troubles. But in fact, he was merely following tradition.

"Teddy! Alexandra!" From across the yard, Mom called down to us from our third-floor window in the big house. "*Obiad!*" Her words bent with her Polish accent as she invited us up for lunch—"*obiad*" in Polish.

Interrupted, Grandma Claire turned around and retreated to her car.

"I just don't understand it," I could hear her mutter to herself as she walked. "It must be all my fault. We sent him to the best schools."

Grandma had been brought up with the notion that a man must have a salaried career.

Unlike Rokeby's aristocratic Chanlers, the Cutlers— Grandma's family on her father's side—had been professional people. Her uncle Elliot Cutler had been a professor of medicine at Harvard and one of the first doctors in the United States to perform open-heart surgery. Another uncle had been chairman of the National Security Council under Eisenhower. Her father had been a senior partner at Smith Barney in New York City from the 1920s until his death in 1950.

I climbed down from my seat of near triumph.

"Coast's clear, Roy," Dad shouted into the barnyard. "You can come out now. Come up to the big house if you want lunch."

I was resentful of Dad's collection of freeloaders, mostly

people of questionable character and sanity, to whom Dad gave more attention than he did to me.

Among the more memorable Rokeby hangers-on were "Bob the Ghost," a schizophrenic; the cryptic, menacing Walter, who according to Dad had robbed graves in Mexico before moving in with us; and Victoria, a diminutive Hungarian lady with dead gray hair and silver teeth, who lived on our living room sofa for a whole year and cried continually into her kerchief for her beloved homeland, rubbed her gnarled hands, and winced from the excruciating pain of her arthritis.

Whenever I'd try to complain about them, Dad would give the same answer.

"My grandmother had boarders living in her house in New York. The idea was that if you had a big house, you should have people staying there. Did I ever tell you about the Palestinian gentleman from Haifa, whose father had been a planter? He lived on the third floor and would kiss my grandmother's hand whenever he saw her. He had some questionable job to do with Jews in Palestine. And the Major? . . . The Major would never kiss my grandmother's hand. . . ."

I tried to keep up with Dad's long strides as we walked up to the big house, passing the coach house, where his pop used to have his machine shop. The thick soles of Dad's construction boots clipped the driveway's hard dirt.

When he was a small kid, after the war, in '46 and '47, Dad used to come up from New York on weekends with his pop. They would work in the machine shop, where Grandpa

Dickie would make parts for machines used on the farm—mostly tractors. He would use nineteenth-century equipment, like a lathe, shaper, grinder, and drill press. Dad told me that the shop used to be very dark, with fifteen-watt bulbs hanging from the ceiling. And it would get very noisy with the flapping of belts and the squeaking and grinding of machines. The building would shake with all the machinery running. These had been happy weekends for Dad.

Nowadays, Aunt Olivia kept her horses and goats there.

CHAPTER FOUR

VENISON STEW

Dad and I entered the forty-foot-long front hall. The ceiling was collapsing in some spots and stained with black mold in others. While spacious, the dark hall was suffocating.

Immediately to the left of the front door was a white marble plaque that read: IN MEMORIAM, STANFORD WHITE, ARCHITECT AND FRIEND. Beside it were two engravings. One was of General George Washington at the entrance of his tent, holding a copy of the Declaration of Independence. In the other, the Marquis de Lafayette stood before his troops after the Battle of Yorktown. Covering the opposite wall was a full-length tapestry depicting Pompey in the Roman Forum. Ribbed with age, it had been scratched and frayed by cats' claws at its bottom edges.

Our kitchen was buried in the center of the first floor, three stories below the rest of our apartment. Unless you went through the dining room to reach it, you had to pass through a dingy, windowless pantry that smelled of leftover cat food cans. Looming over everything was a cupboard where the green-and-gold-rimmed Astor china was stored, to be taken out and used only on rare, ceremonial occasions.

Mom had cleared a section of the long farm table so we could eat, shoving aside a clutter of books, newspapers, and Dad's various grease-stained to-do lists. Over the table hung a ribbon of brown flypaper still plastered with dead flies from the previous summer. The pine floorboards were dirty and almost bare of varnish where they'd been scraped over the years by the legs of our red metal kitchen chairs.

I would frequently look at Grandma Claire's Talbots catalogs, not for the conservative clothing, but for the furnished backgrounds—clean, carpeted floors; neatly set tables with vases full of fresh flowers. And I would try to imagine how different my life would be if my home looked like those furnished spaces. I might even invite school friends over to play.

Mom stood over a lentil-and-venison stew, which boiled

on the stovetop. Her long brown hair was pulled up in a bun that had partially unraveled into wisps around her face. She had a distinctly Slavic beauty, marked by sadness and a seeming passivity. I looked exactly like her.

The venison had been donated by men whom we allowed to hunt at Rokeby. Mom would get Dad to hang any newly slaughtered deer upside down from a water pipe in the old storeroom. She would stand on a ladder in her white canvas apron, machete in hand, and cut away the skin of the deer, busily slicing for hours. I knew from the enormous volume of the antique cauldron that we would be living on this stew for weeks.

"No food until all the grease is cleaned off your hands, Teddy!" Mom was one of Rokeby's fiercest females.

"Yes, dear. Anything you say, dear. . . ." Dad muttered this automatically, without making a move toward any cleaning products or running water.

Mom was a typical Polish wife in many respects. She'd been raised by a Russian mother who, as an infant, had escaped with her aunt from the Bolsheviks to Warsaw. I'd been told stories about how my mother's mother would stay up the entire night to do the family's laundry by hand, or how she would walk up their steep mountain road—about half a mile from the bus stop—carrying heavy groceries. Despite their poverty, Mom's family had had meals on time and clean laundry, china and linen tablecloths for Sunday and holiday meals. In Mom's childhood photos, she and her sister wore neatly ironed dresses and colorful ribbons tied around their freshly shampooed pigtails.

It was Mom who cleaned the front rooms of the big house—particularly the front hall and the three flights of white steps—although, with time, she did this less and less. Hers was a war of one against the mess of generations.

Like Grandma Claire's, Mom's upbringing had also instilled in her an understanding of the importance of money. She'd had jobs over the years—albeit jobs for which she'd been overqualified, with her degree in Russian and German philology from the University of Warsaw. She had picked strawberries, been a carpenter's helper, made mosaics for a public park. Eventually, she'd gone back to school for graphic design and taken a job as a commercial artist.

But in all other ways, Mom had left Poland behind. She now painted her nearly nonexistent eyebrows an electric blue and was very minimally interested in either propriety or family.

"You're a dirty swine!" Mom now kicked Dad's foot in what I had learned to see as a normal expression of spousal affection.

"Ow! Was that really necessary?" Dad laughed.

"Take off this filthy shirt now!" Mom started ripping off Dad's blue work shirt. Buttons went flying.

My unconventional parents were so oddly matched that people could not imagine either of them with anybody else.

I was enchanted by their story. When Dad spent time in

Poland in his twenties, he rented a room from Mom's aunt, "Ciocia Jadzia," and took courses at the University of Warsaw. Mom also lived with her aunt and attended the university. I imagined my parents as young students, staying up late in Warsaw's cafés and strolling through its shadowy streets, trying to steer clear of the secret police. Dad the handsome, Harvard-educated American aristocrat and Mom the meek, hopeful university student.

After they'd married, Dad chose to return to Rokeby, his inheritance, hopeful that he could revive the place. He was soon consumed by the dream and entangled by the reality.

Poland was a place of both physical and spiritual darkness. Mom was full of this darkness. I had it within me as well. I knew that one day I would have to go back to Poland—where Mom had taken me every other winter vacation until I was seven—to reconnect with the darkness that lived inside both of us.

I used to dream of living there with Mom, in her family's roughly stuccoed mountain house, under the protective eye of the sharp-peaked Carpathians. I believed that I would have thrived in a strict Communist system, where being poor was not seen as a mark of a weak character. School, I imagined, would be a place of hard work. The education was based on rote learning, which I loved because facts are never gray.

None of this meant that I wanted to eat like a Polish peasant.

"Is there anything besides venison stew?" I asked as I opened the fridge. Its rusting hinges creaked. It smelled of dead meat. On the first shelf were a pack of hot dogs, a quart of expired milk, a bottle of French's mustard, and a jar of green, furry Ragu. "Can we go to the Tea Garden tonight?" I asked hopefully. A cheap Chinese restaurant called the Tea Garden was the only place we ever went out to eat.

"I can't. Tonight is the last showing of *Fanny and Alexander*."

The movies were Mom's refuge from reality. Occasionally, the arty local theater was our mutual escape. I had received a broad education in foreign and independent film because Mom couldn't afford a babysitter. She had taken me to see *Nosferatu* in black and white, Cocteau's *Orpheus, Women in Love,* and *From Mao to Mozart.*

This last—a documentary about Isaac Stern's visit to China to give master classes to conservatory students there—made me dream of having the discipline to practice my violin like the students in the film. They lived in cubicles and practiced five, eight, ten hours a day.

"Can I come?"

"No. I'm going with a friend who doesn't like children."

Mom was most interested in me when I posed for her sketching sessions.

I stood over the table looking at my dinner. I would gladly have eaten a plate of distinguishable food items neatly arranged—a portion of colorful vegetables, a slab of meat, some rice, like the meals at Grandma Claire's house. I didn't like amorphous brown stews. I decided not to eat.

"I'm not hungry. No stew for me, thanks."

The phone began to ring, as it always did at mealtimes. In addition to his Rokeby groupies, Dad had an extensive network outside Rokeby in the local community. He was usually involved with several charitable missions at any given time. He had a number of immigrant friends whom he was helping obtain refugee status. He would take junk off people's hands to add to his barnyard collection. Or he would offer the use of his backhoe or bulldozer and free labor in exchange for a couple of beers and some ice cream—keeping the barter system alive and well.

Regular correspondents included "Frankie the Freeloader," who used foster children to work his pig farm, and Irving Rothberg. Irving's front lawn was littered with gravestones, as his business was carving messages and biographical information about the dead. Part of this business involved retrieving and transporting corpses and preparing them for burial, so we got to hear plenty of stories about him.

Our telephone—black and square with a rotary dial— sat neglected on its haunches like a fat cat.

"Are you going to answer it?" I asked Mom, who was closest to the phone.

"No, I don't want to," she whined.

We had a party line, shared with both Uncle Harry's family and Grandma Claire, and anyone could listen in. Dad regularly joked that Mom had a fear of telephones, that she was afraid the phone would bite her. But Mom's paranoia of people listening in probably had its source in Communist Poland, where even a harmless exchange over the phone could end in arrest by the secret police, imprisonment, and even disappearance.

I finally answered it. It was one of Dad's Book of the Month clubs. Dad would order from them heavy hardcover books, like encyclopedias and atlases, under the names of various Rokeby pets. Dad's two favorites were "Ms. Mimi Katz"—named after my cat Mimi—and "Mr. Piesek Yaruzelski." This latter subscriber was a bright yellow dog with pointy ears that Grandma had named Yellow Dog Dingo, but whom Dad called Piesek Yaruzelski—"little dog Yaruzelski"—after the last Communist leader of Poland, who imposed martial law in 1981.

"Um . . . Ms. Katz is not here right now. She has gone on vacation," I told the creditor, as Dad had instructed me to do whenever they would call.

After I hung up, Dad and I had a good laugh, while Mom ranted.

"You're both criminals! I hope you end up in prison! Next time they call, I'm going to tell them the truth!"

"Come to think of it, I believe I also got *you* by mail order. But they sent me the wrong sister!" Dad would often joke that Mom had been a mail-order bride, since they'd been married by proxy—a third party had stood in for Dad at the wedding in Poland—so that Mom could leave the country and arrive at Rokeby like a mail delivery.

At that, Mom joined in the laughter. This was rare, as she generally disapproved of Dad and sided with the more powerful family members—his brother, sister-in-law, and mother—who collectively condemned him.

With my parents, I was immersed in a theater of the absurd: the beautiful Polish woman with blue eyebrows and a truculent temper; the filthy gentleman farmer beloved by all, except his closest relatives, for his brilliant mind, generous spirit, and total disregard for public opinion; and their serious young daughter, who mostly acted the part of the parent.

As if on cue, I now walked over to the cabinet above the sink and snatched a pair of nail clippers, then joined Dad at the far end of the table.

"Dad, give me a foot."

Dad absently lifted a leg up. I placed his foot in my lap and began to remove his shoe and sock. He tried to withdraw it. "Now, wait a minute. What are you planning to do with those clippers?"

"Just the usual pedicure. Hold still."

His toenails were thick and yellow like seashells, each with a dense layer of dirt and grease underneath. After clipping the end of each nail, I also dug under it with the metal file and scraped out the black dirt. Dad winced all the while.

"Other foot, please."

"No, no. We can do the other foot some other time."

"No. Now, Dad!"

Someone had to take care of him. Poor Dad sacrificed everything for Rokeby's care and had no time to take care of himself or his family. His gray, rotting teeth, his filthy clothes and skin, his gnarled hair, and the black dirt and oil under his fingernails all cried for my attention.

While I was happy to play the parent, I sometimes fantasized about having overbearing Chinese parents who would help me become as accomplished as the violinists in the Isaac Stern documentary—parents who would furiously scribble notes during my violin lessons and later review them while they supervised my practicing.

As I was finishing up Dad's pedicure, I noticed a pair of shining eyes glaring at me from the gloom beyond the kitchen doorway. It was Aunt Olivia, fixing me with her mad rhinoceros look, her nostrils flared and her eyes fierce.

Uncle Harry was Aunt Olivia's second husband. From her first marriage, she had a teenage son and daughter who were now both away at school.

She'd caught me off guard, amid the mess, with Dad's filthy bare foot in my lap, the cluttered table, the unmatched soup bowls and spoons, the rusty fridge, the dead flies. With her in the room, it all felt shamefully squalid.

"I'd like to speak with Alexandra for a second." Aunt Olivia motioned with her index finger for me to follow her through the pantry and up the back stairs toward her part of the house. I felt light, as if my feet weren't touching the floor and my limbs might detach from my body. Aunt Olivia, an accomplished actress and singer, strutted dramatically.

"We can have our little *conference* in the middle room. Now, come in, and close the door behind you," she ordered.

CHAPTER FIVE

THE CONFERENCE

Aunt Olivia summoned me to the middle room, so called because it was sandwiched between the back and front parts of the house. In the old days, food would be transported from the back kitchen, by way of the dumbwaiter located in the middle room, down half a floor to the old pantry—currently our kitchen—and then out through the swinging door into the dining room. Now the middle room served as the living room in Aunt Olivia and Uncle Harry's part of the house.

Aunt Olivia's figure towered over me like an oversized A, feet planted and hands on hips.

"Well?" Aunt Olivia's nostrils now flared.

I just waited for her to talk. I had nothing to confess.

"I want you to look into my eyes when I speak to you." I tried, but her dark eyes, at times faraway, were now too severe. So I looked at her neck instead. "Now, let's get something very clear." My eyes had already traveled back to the floor. "You are not to call my daughter fat! *Ever!*" She clenched her square jaw. "Do you understand? Do you think that you can bully *my* children?" Like a lioness, she never hesitated when it came to defending her young.

She didn't understand that I was only trying to do what was best for her daughters. As my father did with his protégés, I hoped to mentor and mold my cousins into accomplished and beautiful young women. How could Aunt Olivia feel the need to protect her girls from me when it was I who was protecting them from failure and disappointment?

"You can only play with my daughters if you agree to treat them nicely. No soldiering them around like they're in some kind of military camp. Got it?"

"Yeah, I guess so." I had no voice with her. All my adult confidence and authority had vanished.

"All right, I'm finished," Aunt Olivia concluded, as if she had just given me a perfunctory beating. "You may leave now."

I just kept looking at the floor, my head bowed.

"Are you listening?"

I looked up to see her thin nostrils still flared.

I wished I could ignore things. If I told Mom that a kid was mean to me at school, she would say, "Just ignore them!" But I couldn't. Every critical word pierced me to the quick, an attack on the perfection I worked so hard to cultivate.

"But first, give me a hug." Like a stick, I inched toward her, and she squeezed me for a second. "I really do love you, you know," she said. My eyes were on the doorknob. "You can come into my kitchen now, if you want."

I visited her kitchen often, though it made me uncomfortable. These visits gave me a chance to take note of how Mom's kitchen might be improved if we ever got hold of some money. As Aunt Olivia was also a gourmet cook, I frequently showed up at mealtimes.

As we walked through the bright kitchen's double doors, the smell of cloves and lilacs wafted toward us. The kitchen jutted out the north end of the house, with walls exposed to the east and west. A row of three curtained windows stretched along each of these walls. Leafy green plants hung from ceiling hooks. The hardwood floor had recently been revarnished, and the walls and ceiling were clean, white, and free of cobwebs. On the stove next to the welcoming fireplace sat a gleaming teakettle. The modern refrigerator was covered with reminders, phone numbers, grocery lists, all held up by funny, colorful magnets. Unlike our cabinets, which were stacked with unmatched hand-me-down dishes and glasses, Aunt Olivia's were stacked with matching sets of both. My favorite was a set of blue glasses the color of dusk in winter, what I imagined to be the hue of loneliness. It was here, in this pleasant, well-lit corner of the house, that I felt my position as the poor cousin, poised on the margins of their home life, most acutely.

Maggie and Diana, now back in their jeans, were eating chocolate pudding. They were seated at a slick white linoleum table, uncluttered except by glass bottles of herbs and a crystal vase of lavender lilacs. The girls' chins barely reached the table's surface, despite the fact that they were sitting on Manhattan phone books. I slipped into a chair next to Maggie. "Do you want to go outside and play?" I whispered, keeping my eyes on Aunt Olivia.

"No!" Maggie, imperious, knew she was in control as long as her mother was there.

"Can I have a tiny taste of your pudding?"

"Mo-o-om . . . can Alexandra have a pudding?"

"Ughhh!" she moaned. "Doesn't she have her own food?" Aunt Olivia's dark bun had partially unraveled into wisps around her face. She slapped a pudding down in front of me. "There you are," she said with a sigh.

"Can I take a spoon?" I whispered to Maggie.

"Mo-o-om! Can Alexandra have a spoon?"

"What's the matter with *her* voice today?" Aunt Olivia said mockingly. "The cat got her tongue?" Maggie giggled.

"She can take one herself." Aunt Olivia looked at Maggie as she said this, then giggled too, like a schoolgirl sharing a mean little secret.

I wrestled open the sticky, heavy drawer in the table and picked out a spoon, then swallowed a spoonful of pudding with a hard gulp.

Little blond Diana, oblivious and eager to please, licked her lips happily and pushed the empty plastic cup toward her mother. "More!" She smiled by squeezing her lips together, as if she had been instructed to smile this way.

"Oh, aren't you a regular little piglet!" Being called a piglet was a compliment coming from Aunt Olivia, who adored pigs—clean, theoretical pigs, that is. She had pink pig magnets on her fridge; pink pig oven mitts; and pigs on coffee mugs, notepads, key chains. She even claimed Miss Piggy as a favorite children's show character.

While Aunt Olivia's attention was still on Diana, I took the opportunity to slip out.

DESPERATE TO RETURN to my own part of the house, I rushed back through the middle room. As I climbed the back staircase, my hand slid along the rickety banister. The peeling plaster ceiling loomed overhead like an angry sky. Mounted under the north wall's windows were several sets of horns from Uncle Willie Chanler's hunting expeditions in East Africa. Draped over the horns were cobwebs too high to brush away.

At the top of the back staircase, which was still part of Aunt Olivia and Uncle Harry's territory, I passed through a double doorway. Beyond was a small, shadowy hall—a crevice of the house located so deep within its interior that it rarely got enough light to see by—and a staircase that led up to the third-floor storerooms and our apartment. This dark, dusty hallway, cluttered with broken odds and ends, was the point where Rokeby's three worlds converged: the lonely squalor of the third floor, the elegant formality of the front rooms, and the smug coziness of Aunt Olivia's domain.

CHAPTER SIX

A METICULOUS RECORD

The world of the third floor, my floor, began in earnest at the foot of a narrow staircase. Its contrast with the rest of the house was so great that I felt ashamed each time I climbed these stairs.

The glare of the bare bulb overhead highlighted the peeling white paint on the steps. The pink wall and white banister had been seasoned over the years by the grime of passing hands.

At the top of the stairs, standing at the north end of the central light well, I could discern Uncle Harry through the dusty interior windows. He stood in an alcove among Great-Grandma Margaret's old steamer trunks. Although he lovingly refolded the gowns Maggie and Diana had been playing in, he wore an agitated expression, clearly distressed that the dresses had been disturbed.

Around his stooped figure stood the silhouettes of dress mannequins, rocking horses, slightly broken Victorian dolls, children's desks at which the orphans had been homeschooled. I would sometimes go back there to ride the creaking rocking horses and look through the antique toys, but there was something distinctly unsatisfying about playing with broken toys from another era. Pieces of them were usually missing or would fall off in my hands.

Most of the third floor consisted of padlocked storage rooms where the family archives were kept. Though these technically belonged to the whole family, Uncle Harry kept most of them locked away from the rest of us, as if he were the family's sole true heir.

I only got an occasional glimpse of what lay inside these mysterious storerooms. I had spied bookshelves lined up library-style, with stacks of documents—in and out of boxes—in the aisles, yellowing posters, medals, broken chairs, old mattresses, and stuffed animal heads on the walls.

Everything was covered with pieces of the crumbling, water-damaged ceiling. Metal pails were poised under the known leaky spots. Here, among this mess, lay our history, an addendum to the museum downstairs.

This should have been our part of the house, exclusively. Uncle Harry had seven spacious rooms in his part—none of which were used for general storage by other family members—while my family kept only three small rooms for ourselves. I liked to imagine the third-floor storerooms reclaimed, cleaned, and renovated. Then my parents could have their own guest rooms and be able to invite people to the house whenever they liked, without having to get approval from the extended family.

Uncle Harry closed the trunk and brushed off some of its dust as he took a cursory glance around the room. In his role as guardian of our inheritance, my uncle knew exactly what we owned and exactly where it all was. About once a month, late at night, he used a flashlight to take a full inventory of all the objects in the house: books, vases, lamps, portraits, and items in third-floor storage. Sometimes a spooked guest would report the sound of footsteps during the night.

"It's only Uncle Harry, checking the house for theft."

Now, done with his inspection, Uncle Harry began walking in my direction. As he passed me on his way downstairs, I melted into the shadows of boxes so he wouldn't notice me. But it wasn't necessary, as his mind often dwelled in the past perfect, making him oblivious to the present.

Behind me was the door to the shaft of an old-fashioned hand-operated elevator, which extended behind our kitchen on the first floor. Within minutes, Uncle Harry's voice rose through the shaft. I already knew the text.

"Why is it that I must cover your share of the taxes each and every pay period?!"

As the rent from Rokeby's outbuildings didn't cover the estate's considerable property taxes, Uncle Harry would come into our kitchen and roar at Dad for not being able to come up with his share of the tax money. He'd threaten to confiscate Dad's shares because of his lack of contribution to Rokeby. Finally, Uncle Harry would reluctantly agree to cover for Dad, but he would let it be known that he was keeping a meticulous record of the debt.

It was this tax debt that turned Dad into a willing slave, not demanding any compensation for his physical maintenance of Rokeby.

It was always in the wake of Uncle Harry's tirades about the taxes that I could sense Dad's heavy despair over his inability to find a regular job and earn money. He had worked once as a hospital orderly in New York City during a year off from Harvard. But he had found it difficult to stick to an arbitrary schedule imposed on him by others, and Rokeby continually called to him, like a kingdom missing its king.

Dad would never fight back. Unlike Uncle Harry, he didn't keep a meticulous record of his contributions to Rokeby.

He would just stand uneasily, one foot slightly in front of the other, rubbing his head and shrugging.

"Well . . . I'll pay you back as soon as I can."

"Hmpph!" I could almost see Uncle Harry laughing scornfully. "You've never worked a day in your life. How will you ever be able to repay me?"

As part of his lifelong commitment to keeping Rokeby in the family, Uncle Harry would also lecture us, the next generation, on the importance of making it in the "real world."

"Each of you must learn a profession. One day, you will have to pay the Rokeby taxes out of your own pockets. You can't expect Rokeby to support you. There are no more multimillion-dollar trusts waiting to open when you reach your majority."

Whenever he said this, I wanted to remind him that if we sold Rokeby, we could each have a very comfortable life. Nevertheless, I intended to take his advice.

I meant to learn a profession. But for me, learning a profession and earning a living would be a way not to keep Rokeby—but to leave it.

🌾

The first thing you saw as you entered our living room was a small wool tapestry hanging on the wall. It was Mom's portrait of our nuclear family. My two ponytails were strings of brown yarn hanging down off the two-dimensional surface, my shoes were laced with yarn, and Dad had a bunch of golden-brown yarn on each side of his balding crown, snaking in and out of the surface to re-create the effect of his hair's wiry texture.

Despite its peeling floor paint; mismatched, broken furniture; and bare bulbs, our apartment was a refuge from the open expanses of the rest of the house. Here I could close, and even lock, the doors. The ceilings weren't high and the rooms weren't sprawling. There were no reminders of the past. None of the extended family—aside from Maggie and Diana, who were my often welcome playmates—ever set foot here. In our apartment, it was possible to maintain an identity apart from our ancestors.

There were no photos or pictures on my plain white plaster walls, no sentimental family artifacts. My ideal living space was a cubicle, so I did my best to keep my room simple. I had basic furniture—a primitive metal cot, a bureau whose drawers had sharp screw ends where the knobs were missing. Against one wall was a five-foot-tall Victorian dollhouse, my prized possession, even though it was not officially "mine" but a Rokeby heirloom. It stood on stilts and had four open-faced chambers, two on each floor, connected by archways. In the middle of the room stood my music stand, with my music book open to the Vivaldi A-minor violin concerto, which I would soon be performing at my end-of-the-year student recital.

Hanging on my bedroom wall was my daily schedule,

which I checked religiously. I'd written it for myself when I was seven. It read:

What I Do All Day:

1. Wake up at 6:30.
2. Brush teeth and hair.
3. Get dressed.
4. Practice clarinet, 7–7:20.
5. Watch *Captain Kangaroo*.
6. Eat breakfast.
7. Walk or ride bike to bus stop, depending on the weather.
8. Return from school.
9. Have snack.
10. Watch *Little House on the Prairie*, 4–5.
11. Practice violin, 5:15–6:15.
12. Have dinner.
13. Practice piano, 6:45–7:15.
14. Do homework.
15. Read in bed.
16. Write in diary.
17. Say prayers.

And that's my day.

I hoped that this list might someday serve as evidence to future generations of my disciplined, serious character.

I often took out my old math tests and English essays.

All A's, high nineties or one hundreds. I ran my fingers over the gold stars stuck on the top of each page and savored Mrs. Keaton's comments: "Excellent" and "Outstanding!"

Everything about Mrs. Keaton, my fifth-grade teacher, was straight and square. She seemed to have a metal pole for a backbone. Both rows of teeth between her canines were in a perfectly straight line, making her jaw seem clenched. She would march her class through the school halls in her square-heeled shoes like a general leading a children's crusade. Mrs. Keaton adored me, as I did everything exactly as she demanded. And I adored her, as she was the only authority figure in my life who was both powerful and fair.

Still smarting from my "conference" with Aunt Olivia, I picked up my violin and began to play madly, as loudly as I could. I wanted my aunt to hear how well I played. I wanted her to know about my gold stars. I would force on her my best self, which she could not seem to see.

My eyes rested on my dolls, displayed in a long row on an antique daybed by the window. But instead of my usual nurturing feeling, I felt nothing for these, my babies. My mind was already leaving them, scrambling furiously into the future.

As I played on, I imagined myself onstage, a famous performer in front of thousands, with an orchestra behind me. Fame meant recognition. Perhaps fame could rescue me from the confusion and shame of Rokeby.

One day—after I'd gone to Juilliard and become rich

and famous—I'd return to Rokeby in a black limousine. All of the inhabitants of the big house would rush out to greet me and kiss my hands in gratitude for having donated millions of dollars to Rokeby's restoration. The barnyard would be clean, the front rooms scrubbed by a paid housemaid, the storerooms converted into private rooms my parents could call their own, and the archives—our joint heritage—organized and available for all to share.

PART II

ELEMENTS
OF DISORDER

CHAPTER SEVEN

SUNDAY MORNINGS

Rokeby had been passed down through the female line for generations. John Armstrong Jr. built the big house on his wife's Livingston land. After marrying the Armstrongs' daughter, William B. Astor likewise moved to Rokeby. The Astors' granddaughter Maddie Ward Chanler inherited Rokeby, and eventually her daughter, my great-grandma Margaret, became Rokeby's sole owner.

Grandma Claire, who married Great-Grandma Margaret's son, was the family's current matriarch.

It was Sunday morning. As I entered Grandma Claire's house through the kitchen, I stepped over cracked linoleum and shivered with disgust at the layer of hardened white grease in the cast-iron frying pan and the mouse droppings sprinkled about.

Grandma had never learned how to keep house. Despite the fact that she entertained frequently, her house was always messy. Books—some from the library and others from her own musty shelves—were piled on the reading table beside her recliner. Though wooden chairs lay on the sofas to protect them from white dog hairs, hair littered the floor and the ragged Oriental rug in front of the fireplace.

Grandma's father had made a fortune as a senior partner at Smith Barney in the 1920s. Before his money had been swept away in the Great Depression, Grandma had been ashamed of her family's extravagant wealth. As a schoolgirl, she would ask her father's chauffeur to drop her off a block away from her school, embarrassed to be seen by the other girls waiting out in the schoolyard. For her, the extravagance of her father's Rolls-Royce—gleaming, self-important, chauffeur driven—had clashed with her own humility; she was tentative and withdrawn.

This was why Grandma—who lived off the interest from some mysterious surviving trust—never hired help, except when she paid her granddaughters to help out at dinner parties.

"Good morning, dear." Grandma Claire was seated at her dining table in her maroon felt bathrobe, reading the Sunday *Times*. "Can I get you some breakfast?"

Grandma's house smelled of musty antiques. It was full of simple, elegant objects: silver ashtrays; glass candlesticks; watercolors of Florence, her favorite city; and a lazy Susan in the middle of her oval oak dining table that held quaint ceramic butter dishes and salt and pepper shakers. There were blue and white Dutch tiles around the fireplace. The gilt mirrors were dim, and the Oriental rugs tattered. Her bookshelves were lined with political biographies and illustrated animal books. Her living room cupboard was full of old parlor games, like backgammon, Scrabble, and Anagrams, all with crumbling boards and missing or cracked pieces.

The long living room was lined with French doors that opened onto a backyard. The dining table was at the east end of the giant room, and sofas were arranged around the fireplace at its west end.

"Are you coming to church with me?" Grandma asked. I usually went to church with Grandma. Occasionally Maggie and Diana would join us, but more often they would go with their mother.

Very rarely, Mom would come to church. She had technically converted to Episcopalianism from Catholicism when she married Dad, although she was really what Dad called a "pagan." She practiced rites like dancing around bonfires on the summer solstice and burning candles for the "spirits."

Dad always claimed to be a devout Episcopalian. When asked about his religious beliefs, he'd say, "They're listed in the back of the Anglican prayer book." But I only ever saw Dad in church on Christmas Eve.

In religious matters, Uncle Harry's loyalty remained with Christ Church in the village, the construction of which had been funded by William B. Astor in the 1850s, because he and his wife wished to attend Anglican services. The Astor orphans had likewise worshiped at Christ Church every Sunday.

Grandma had asserted her independence by choosing a different church: St. John the Evangelist, a quaint bluish-gray Carpenter Gothic with a matching rectory that was just a mile up the road.

At church, we knew all the regular congregants and were related to several. This positive social interaction enticed me to attend regularly.

A second enticement was Sunday-morning breakfast with Grandma Claire, which was a welcome relief from the hard black bread of home. She would serve me sliced grapefruit, and buttered toast, and eggs that she would cook sunny-side up, special for me.

Breakfast done, I'd accompany Grandma Claire to her room to help her get ready for church.

Grandma and I would dress up, transforming ourselves

into unquestionable respectability, she in her black dress suit and I in one of the kilts she'd bought that made me look like a small Talbots model.

Clothes were draped over the footboard of Grandma Claire's bed, and unmatched slippers and shoes were scattered underneath. As she was constantly writing letters and giving gifts, her two dressers were piled with envelopes, books of stamps, stationery boxes, and rolls of wrapping paper and scissors.

Grandma Claire now half sat, half lay on her bed to get into her stockings. Chronically overwhelmed and disorganized, she would coax herself through the steps of any task. "Ughh! Oh dear me! All right now. Right foot's in." She managed to get her right stocking on without tearing it on her curling yellow toenails. "Ugh!" She pulled it high and snapped it to her girdle: *Click.* "There! Now, that's *much* better!"

When she'd snapped both stockings on, she buttoned herself into her inherited, threadbare Lord and Taylor suit, which didn't quite fit her tall, stooped, emaciated body. The jacket's sleeves weren't long enough and left her skinny wrists and hands, slightly twisted with arthritis, to jut out like the front feet of a mole.

Then the search for the shoes began. She bent over to peer under the bed and rummaged through her musty closet. "Ugh! I only see one black loafer. *Who* on earth could have taken my other loafer? Ughh!" Then, "Oh I really *am* a dunce. How *could* I have misplaced my glasses?"

"They're hanging around your neck, Grandma."

"Oh, for goodness' sake! . . . *Where is my pocketbook?* I just *can't* go anywhere without my pocketbook! I'll give you ten dollars if you can find it."

Things went on in this way until we got into the lemon-yellow Plymouth and went floating along Rokeby's dusty farm road as I stared out at the open fields.

My reverie was broken by a noisy red Fiat bumbling toward us, undaunted by the driveway's potholes. Grandma Claire began pumping her foot, madly alternating between the gas and the brake, jolting the car back and forth. I thought I could hear her teeth grinding and the short, snorting breaths from her nose, building like a dragon's internal bellows. When the Fiat reached us, Grandma Claire came to a complete stop and rolled open her window with uncharacteristic speed.

The red car stopped obediently alongside us, like a child caught in the middle of an exciting game of tag. The female driver cheerfully rolled down her window to greet us.

"*Bonjour,* Madame Claire, *comment ça va?*" Her girlish voice rang like a soft bell. A bit of perfume wafted from her direction, and the gold bangles she wore clinked against the steering wheel. Like me, she had a gap between her front teeth.

A roar welled up from deep inside Grandma Claire, and, like lightning, she slammed her car door several times against the red body of the Fiat:

Once! "You French harlot!" Grandma Claire's sixty-six-year-old voice shrieked over the din of colliding metal.

Twice! "Get off my property!"

Three times! "I'll smash you!"

Four times! "Get back to France!"

The red car pulled away before the fifth blow. Grandma Claire now had one leg out of the Plymouth and was shaking her fist, still screaming.

Grandma's violent reactions to Dad's friends and general way of life were a testament to how desperately she cared, how desperately she hoped to set Dad straight. The company he kept reminded her painfully of her failure to raise him into a model member of the class into which he'd been born.

"What did she do, Grandma?" I asked.

"Oh . . . well . . ." She seemed disoriented now, her anger suddenly gone. "She's a *fallen woman*—married with three daughters—and a *homewrecker*."

The only wrecks I had known of until now were the junked cars Dad had accumulated in the barnyard.

And Mom often said that Dad had wrecked her life.

And now this woman's car door was wrecked, as an act of revenge: one wreck for another.

GRANDMA SEEMED TO have recovered from our encounter with the harlot by the time we arrived, late as usual, to the ten o'clock service.

We shuffled in along the red carpet—*the blood of Christ*—that led from the back entrance all the way up to the parapet where congregants took Communion. It was dim in the sanctuary, with its stained glass of dark purples and blues and its dark brown pews. To keep from falling asleep, I would scrutinize the pictures in the glass.

"Ugh, dear," Grandma Claire murmured. Her skirt and slip made a rustling sound as she slid down onto the prayer bench. She rested her forehead in her skeletal hand. I was touched by the way she communed with God on her knees, while the rest of the congregation was somewhere else in the prayer service. I assumed she was apologizing, as she tended to blame herself for the family's problems.

I usually communicated with God in writing—anxious pleas in my journals for a better life, for a clean and orderly home, for acknowledgment from the people around me. I believed that God watched my every move and noted my hard work.

I was not at all satisfied with Christianity itself, however, and only really came to church for the social benefits. I felt that religion should be more than learning campy songs in Sunday school and weekly attendance at church. I wanted to be commanded in much more certain terms.

During Communion, Grandma Claire always remained in her seat, because alcohol was used—even though the minister said it was blood. Grandma preferred to drink at home, alone in her bathroom, from green, fish-shaped bottles.

After the service, Grandma Claire and I walked up the hill to the cemetery behind the church, to visit her dead husband, Grandpa Dickie. The sun brought droplets of sweat to Grandma's face as she drifted off into the past in front of the simple gray slab, speaking as if she were making a confession to me.

"I took the kids away from Dickie and Rokeby when his drinking had become impossible," she said as she cast her eyes downward apologetically, "and my mother-in-law made it quite clear that Rokeby was *her* property. All we had there was the tower bedroom. . . . So I decided to stay in New York with the children. Liz went to Chapin, and Harry to Buckley. Your father, of course, was a problem child. Sending him off to boarding school seemed like the right thing to do at the time. . . . We'd visit Rokeby in summers, and sometimes for holidays, of course. . . ." I wondered if she regretted taking the children away from their father, who had lived with his mother at Rokeby for the final decade of his life.

Then Grandma Claire remembered herself and patted the stone. "Everybody loved your grandpa."

I prayed silently that Grandma Claire would not die any time soon. She was the person who paid for and drove me to my violin lessons, bought my school supplies and clothes, and had shelves stocked with food when our own supplies were depleted. Also, from time to time, she would mention the possibility of sending me to board with our cousin Lilly on the

Upper East Side so that I might attend Chapin, her alma mater. I clung to Grandma Claire as if to a life raft that both kept me afloat in the present and would steer me out of Rokeby in the future.

Back at Rokeby after the service, the lemon-colored Plymouth sailed on past Grandma Claire's house, past the redbrick coach house, and pulled up alongside the grassy triangle in the center of the barnyard.

Grandma Claire slowed the car as she scanned the barnyard with her uncanny vision for all things Dad. Through the car's open windows, we could hear the roar of his backhoe from down the hill, by the lower barn.

"*Now* what's that father of yours up to?"

Grandma Claire parked and began making her way down the hill toward him. I followed her.

Dad was digging a trench.

The backhoe seat had been twisted to face the rear of the machine. Dad's hands were on the levers. The backhoe's feet were planted on the ground to each side, to steady the center while the hoe—like a giant squid arm—flailed out and around to the side of the trench. Then its hand opened and dumped more soil onto a pile at the edge of the gaping hole it had created.

Grandma Claire was waving her arms, signaling for Dad

to shut off the engine. He did, then remained in the backhoe seat for the anticipated scolding.

"Now, Teddy! I don't want that French creature anywhere near this property or this family! Do you understand, Teddy? You have a wife and child, for goodness' sake!"

"I understand," Dad responded automatically.

"And dinner tonight is at my house, at seven." Grandma Claire's resentments never interfered with her hospitality.

CHAPTER EIGHT

THE OUTLAW

Late that afternoon, I was alone in our kitchen when Dad entered with a load of TV dinners—rejects from the local pie factory. Dad had made friends with the owner of the factory and would sometimes get the dented TV dinners, as well as the leftover pie dough. He used the latter to feed his giant pig, Egbert.

On his heels was the French woman Grandma had assaulted on our way to church. Now she loped along, glancing over her shoulder as if pursued.

"Brought you a surprise!" Dad said ambiguously as he piled the TV dinners on the table and ushered in this latest stranger.

I marveled at the pile of aluminum trays: Instant dinner! A miraculous invention!

Without another word, Dad picked up the phone and made a call, leaving the introductions to me and the French woman.

"Yeah, Rick! I got those two-by-fours you wanted. They're in the barnyard. . . . What's new with the cleanup down at the tracks?" A freight train had jumped the tracks, and various local residents had been scavenging the spilled contents.

"You must be Alexandra!" the woman said to me in a seductive French accent. "I've heard so much about you! I'm Giselle." Her eyes darted around as she spoke.

I was confused. Grandma Claire's reaction to Giselle earlier that day made me suspicious, as did Giselle's furtive manner. It was clear even to me that she was not a legitimate family friend, or even a mutual friend of my parents. On the other hand, she was rather too fashionably dressed to be one of Dad's usual "riffraff"—as Grandma Claire called his protégés. She wore a brown leather jacket, cracked with age, and

Levi's. Even if she had been the usual type, I was well aware that both Grandma and Uncle Harry tended to overreact to Dad's friends. Dad referred to his mother and brother as the "fun police," claiming that the only reason the family kept such hard, lumpy mattresses in the guest rooms was to discourage guests from staying.

So in my confusion I assumed no attitude at all toward Giselle, neither respectful nor hostile — just blank.

However, the moment I let her take my hand in hers— her gold wrist bangles jingling—and cajole a disinterested smile out of me, I regretted that I hadn't decisively discouraged her with a scowl. At ease now that she felt she'd won me over, Giselle took a seat at our kitchen table and began running her hand through her attractively disheveled hair.

Fighting a sense that I'd betrayed Mom and hoping to get Giselle out of her kitchen, I turned to Dad.

"Hey, thanks for the TV dinners, but we still need to go food shopping. The fridge is empty."

"Just a minute . . . Let me see." Dad dug into his shirt pocket and pulled out a wad of folded and soiled papers, mostly to-do lists written on the backs of bills. He rubbed them between the greasy thumb and forefinger of one hand to separate them, then tossed the papers onto the table. "See what you can find there."

I saw a green edge sticking out from the middle of the pile, but it turned out to be only a ten-dollar bill. I knew we'd be paying a visit to Yishmael at the Mobil station.

Dad and I were headed to the cab of his green Chevy when Giselle shot past me like a small child who wanted the best seat. She wedged herself behind the stick shift, among old newspapers, bottles of motor oil and antifreeze, a half-eaten sandwich, and empty soda cans, and nestled next to Dad, who had just gotten in on the other side.

I sat in the cab, squeezed up against Giselle, embarrassed. I'd never seen Dad spend time with a woman in this way. Neither she nor Dad had offered me any explanation of who she was. Neither acted as though it was unusual for a complete stranger to go grocery shopping with us.

Giselle aside, it was a familiar ritual: Dad would brave the county road, tempting fate with his expired inspection sticker, broken taillight, and cracked windshield, to drive to Yishmael's Mobil station so we could borrow money for groceries.

When I was little, I would stick my head out the cab window and let the wind slap my face and suck my long, flapping hair back as Dad raced along this strip of highway, usually with an open beer bottle between his knees, just under the sight line of any passing cops.

We found Yishmael in his office.

"Ooooohhh . . ." He laughed, delighted to see us, his smile revealing his silver front teeth. He got up, making a hand sandwich with Dad's right hand between his two. Then he shook my hand, his skin rough and dry, stretchy and very dark.

"Yes! Young daughter!" His *r*'s rolled in his thick Turkish accent. "So pretty now! How old?"

"Ten."

Yishmael suddenly raised his eyebrows in bemusement as he noticed Giselle standing outside by the truck. She stood with her hands in her pockets, looking at her feet. She kicked at broken pieces of asphalt, her hair around her face as though she were the bored, restless child waiting for Dad.

I quickly turned my attention back to Yishmael and smiled politely, hoping to expedite the formalities. Yishmael knew why we were here. Yet, when one was coming to ask for a loan, when one had not paid one's previous loans, perhaps ever, one had to discuss whatever one's creditor wished to discuss.

"How is wife?"

"Oh, having a bad day, I would say. Wouldn't you say so, Alexandra?"

"Yeah, I guess."

"Still doing the, what you call . . ." Yishmael made knitting motions with his hands.

"Knitting? She doesn't knit, does she?"

"He means weaving, Dad."

"Actually, you're asking the wrong person. I don't keep track of *what* she does. I can't be around her for more than a few minutes before she goes on the offensive."

I tried to catch Dad's eye and give him a significant look, but he eluded me.

"Well, we've got to move on. Hey, Yishmael, do you think I could borrow another fifty? The kid needs to eat again."

"How much? Fifty? Sure."

"Just be sure to note it down." Dad made it sound as if he were serious about paying it back.

"Yeah, yeah, it's all in the book."

Dad and Giselle dropped me off at the A & P, with my calculator and my shopping list: Lender's onion bagels, whipped cream cheese, eggs, frozen orange juice concentrate, hamburger meat, spaghetti, Lipton chicken noodle soup packs, milk, twenty Fancy Feast cat food cans, a large bag of generic dried cat food, and a pair of No Nonsense socks. To me, having clean white socks was a necessity, even though they cost four dollars out of my sixty.

I found it comforting to push a shopping cart up and down aisles full of food. Muzak was playing in the background, and the men who stocked the shelves always seemed to be whistling.

Yet I felt envious of the women who could afford to fill their carts to overflowing. The items I could buy with sixty dollars just barely covered the bottom of my wagon. And it was always nerve-racking to stand by as the cashier rang up my items. I made sure no one I knew was nearby, in case I ran over budget and had to return things.

Having paid without incident, I went outside to find the Chevy, but it wasn't there. As I waited in the parking lot for Dad and Giselle, people rattled past with shopping carts, their shadows lengthening in the twilight. Car doors opened and closed. The evening air carried some cold patches, bringing goose pimples to my bare arms and legs.

It was getting late, and I began to feel self-conscious. I froze my face into its usual confident expression in order to discourage any concerned adults from asking me if I needed a ride, or if I needed them to call anyone for me.

When the Chevy did finally appear, I was relieved to see that Dad was alone, but not thrilled when he told me he still had some stops to make on our way home.

"I want to check up on the leftovers from the train wreck."

To get to the railroad tracks that ran along the edge of the Hudson River, we had to drive through the heart of the cozy hamlet of Barrytown, just north of Rokeby. It had its own zip code and its own one-room post office, which had been built to look like a Greek temple. Aside from four estates, the hamlet consisted of single-family houses scattered up and down two extremely steep hills that descended to the railroad tracks and the river. Many of these houses had belonged, at one time, to Aunt Elizabeth, Great-Grandma Margaret's sister.

On the other side of the tracks was a grand house that sat on a small piece of land, which included the post office—so one could say that the gentleman who owned this property also owned our zip code. He was a friend of Uncle Harry's, as he shared his passion for historic preservation. I'd never actually seen the current owner, as he owned many grand houses and chose not to stay in this one for any length of time. But his presence was felt strongly enough that nobody dared cross the tracks to his entrance gate. We could see his house—with its weeping willows draping the front lawn—from the north end of Astor Point, a rocky riverfront outcropping at Rokeby's western edge. His house was not like ours. He had not inherited it, but rather purchased it from Gore Vidal—who had bought it in the 1940s—and done a perfect historic renovation.

I walked along the tracks behind Dad as he scanned the ground with a flashlight. "Looks like the place has been picked clean. . . ." Dad had already been down here several times, and taken loads of plywood and cases of Schaefer beer that had been spewed from the derailed freight train.

"Warren Delano was killed in a train accident on these tracks," Dad mused. Warren Delano—FDR's uncle and president of the Delano Coal Company—had lived on the estate that bordered Rokeby to the south. "It was 1920. Warren Delano was on his way down to the post office to pick up the mail with a four-in-hand, and the horses bolted, as a 'twentieth century'—a crack, express passenger train—was coming through at about ninety miles an hour, with no stop in Barrytown. It would go from New York to Chicago in seventeen hours. . . . It didn't help that Delano was a famous driver successful in racing competitions. He couldn't control those horses once they'd panicked and bolted. When they ran right into that train, all four horses, together with Warren, were caught in the wheels and dragged along. . . . The irony was that Warren Delano was on the board of directors of the Atlantic

Coast Line consolidated railroad that ran from north Virginia to Florida."

"Dad?" I asked anxiously. "Can you get arrested for taking stuff off the tracks?"

"Me, arrested? Ha!"

CHAPTER NINE

PAST PERFECT

VANITY FAIR.   May 18, 1872.

No. 183.   STATESMEN, No. 112.
                          "Consequential Damages."

O h good evening, dear . . . *hiccup* . . . You're jus' in time
to help with the dinner party. . . ." Grandma Claire wel-
comed me as I rushed in through her kitchen door, after having

unpacked our groceries up at the big house. "Whaaaa . . . ?"
she asked guiltily, in reaction to my scowl.

She was oscillating from side to side by the stove, as if unable to choose which way to fall. As she turned unsteadily to
face me, I saw that her eyes had that milky, sleepy look, with
her lids half closed. She seemed to grin through me. Her nose
was red and swollen, in contrast to the deathly whiteness of
her complexion. Because her posture had been permanently
lost to a hunched back, her rocking gave her the aspect of a
wrestler preparing to take me down.

On her mother's side, Grandma Claire was descended
from the Fishes, a very old New York family founded by an
Englishman named Preserved Fish. Born in the 1760s, he'd
been thus named because as a boy he'd survived a shipwreck
and been plucked from the sea while crossing the Atlantic
from England. Grandma Claire's great-great-grandfather
Nicholas Fish had been Alexander Hamilton's fellow student
at King's College—what is now Columbia University—and
had served as an executor of Hamilton's estate. As a result,
Nicholas named his son, who was born shortly afterward,
Hamilton. He was the first of many generations of Fishes who
were politicians named Hamilton. Hamilton Fish I served as
secretary of state under Ulysses S. Grant. His grandson was
Grandma Claire's uncle Hamilton Fish III—known to us as
"Uncle Ham"—who served in the U.S. Congress as a Repub-
lican from 1920 until 1945, and was chairman of the House

Foreign Affairs Committee and one of America's leading isola-
tionists throughout the 1930s.

But despite her distinguished— if in the latter case mor-
ally dubious—lineage, and the fact that she still had enough
money to live comfortably and help with her grandchildren,
Grandma Claire's life at Rokeby was difficult. The setting,
with its long, dusty driveways riddled with potholes, was rus-
tic. And Dad, whom she wanted so desperately to control,
broke every rule of propriety that she held sacred.

So to help ease the stress, Grandma Claire drank.
Though the drinking made her prone to sudden rages, I
would most often find her asleep in her bed when I stopped
by after school, her face pale and drained of life, her nose red
as if she'd been crying. She'd open her filmy eyes and hiccup
constantly in alarmed little gasps.

Grandma Claire had moved permanently to Rokeby—
more affordable than renting an apartment on Manhattan's
Upper East Side—once Aunt Liz had finished Chapin and gone
off to college. With Great-Grandma Margaret and Grandpa
Dickie dead, Grandma Claire had become one of Rokeby's
owners. As such, she was drawn to the property, like a concerned parent.

Grandma Claire now pointed to boxes of Triscuits and
Wheat Thins and a chunk of cheddar cheese. "Uhhh . . ."
She moaned in confusion as she took an unsteady step. "Oh
dear!" Her movements were slow, her nasal breathing loud.

"Could you take these out? . . . I'll give you twenty dollars later." She frequently slipped us kids payment for helping out at her dinner parties. We would set and clear the table, pass the hors d'oeuvres, and wash the dishes. It was tacitly understood that the money would also ensure our good manners. *Be sure to curtsy and smile.*

I was too angry at her for "getting bombed" before a dinner party to give my usual modest refusal: *You don't have to.* I always ended up accepting the money anyway.

The first thing you saw as you entered Grandma's main room was the bar. Set up on a narrow table against the back of the orange "dog sofa," it was well stocked with Ocean Spray cranberry juice, Clamato, spicy V8, gin, bourbon, and pink sherry in a glass decanter.

Maggie and Diana played jacks on the floor in the recreation area—complete with a black-and-white TV—behind the sofas where sat Mom, Aunt Olivia, Uncle Harry, and an older couple who'd been family friends for decades.

Dad hadn't yet appeared, but I knew he would because he never turned down a free meal.

With Grandma in such a state, I hurried back into the kitchen, now steamed up by the boiling pots.

Grandma Claire was wearing a white canvas apron that said COMPLIMENTS TO THE CHEF and sported black burn marks along some of its edges.

I grabbed her oven mitts and removed the leg of lamb

from the oven, placing it on the stovetop next to a pot of boiling artichokes, a dish of baked scallops, and a pan of sautéed asparagus. The scent of rosemary rose from its sizzling juices. Grandma could prepare these dishes in her sleep, or in any other condition.

The table was set with custom-made place mats, which bore photos from Grandma's youth and childhood—now distorted by years of spills. Some photos were of the grandchildren, and others were of Dad, Uncle Harry, and their sister, Aunt Liz, as kids. There were still others, black-and-white photos, of a young Grandma Claire, her four siblings, and their parents. The blank spots in the mats were like the ghosts of banished relatives.

GRANDMA WAS AN expert at seating arrangements. "Now, Ala"—her voice sounded hollow and slightly breathless— "you're there—*hiccup*—next to me." Grandma was very protective of Mom. She would refer to her as "poor Ala" and keep her close by, under her wing. "And Olivia—*hiccup*—I hope you can—*hiccup*—forgive me. I've squeezed you in between your husband and your brother-in-law."

"Oh, that's just *fab*ulous, Claire! We can talk about the Porcellian Club and Haybines!" she muttered through a crooked smirk, but her sarcasm was lost on Grandma.

"And Teddy is at the head of the table. But oh, now, where

is Teddy, damn it all?" Grandma always sat at one end of the table, while Dad, being the oldest, got the honor of sitting at the other.

Just then, Dad rushed in with bits of hay still in his hair. He apologized for his appearance. " . . . Just didn't get a chance to change."

"Oh, no need to explain yourself. It's not as if anyone here doesn't *know* you," Aunt Olivia said, rolling her eyes.

Amid the general clatter of knives and forks against china, the conversation revolved around the least controversial subject—the past. Dad took the lead with stories of his outrageously naughty childhood.

Dad hardly ever asserted his opinions in front of his siblings or his mother. If he did, or if he received any positive attention from other people for his wit and charm, his immediate family would mock or contradict him. It was easier to avoid humiliation by keeping quiet or talking about noncontroversial matters, and on the subject of his childhood hijinks, everyone could agree.

Once, at the age of four, during a lunch party at his great-aunt's, Dad had locked himself in the bathroom and flushed the key down the toilet.

"I was trying to figure out a way to climb out the window—which had no ledge—and get back in through another window. I was going to try to grab the next window and swing over. There were hasty consultations audible from behind the door as to what could possibly be done. It was a really solid door, about two inches thick, and there was no easy way to get through it. Finally the superintendent was called. It was a Sunday afternoon, so he came with ill grace to take the door down. He pulled the pins, and pried the door off the hinge side, and got it out."

Aunt Olivia erupted into histrionic, though not un-friendly, laughter. "Oh *hon*estly! I don't believe you were *that* awful!"

"Oh, he was. . . ." Grandma sighed, her cloudy eyes still distant. Her body was curled up in her seat, her shoulders stooped, her abdomen caved in. Every part of her was gnarled.

"Oh *do* tell us another," Aunt Olivia teased.

Dad told about the time when, at the age of four or five, he had been dragged along to his uncle's wedding re-ception at an incredibly narrow Manhattan brownstone. He had grabbed handfuls of brightly colored canapés that he'd assumed were pastries, and was disappointed to discover that they were actually salt fish.

"There was nowhere to put said canapés, so I just jammed them into my pants and jacket pockets. There were quite a lot of them, and with the pressure of people pushing up against me, this stuff started to ooze out, and naturally got onto other people's clothes, which they began to notice. People were starting to get mad at me, and chase after me, at which point I had to escape. So I snuck upstairs. But since the place was so full of people—even upstairs—the only way I could go was out. So I climbed out onto a cornice. At some point, somebody noticed me up there, and a commotion erupted on the street, while from inside there was this consternation among the guests, who were trying to get me back inside. But since nobody wanted to climb out onto that gutter to get me, I sat pretty up there until eventually I had to come in because I wanted something to eat. At that point, I got into serious trouble and was taken home. And that was the end of the party!"

"Speaking of delinquency," Uncle Harry interrupted. "When do you plan on replacing the roof on the pump house?"

Grandma Claire bared her large teeth, ready to join in the attack on Dad. *You, Teddy, are a disappointment, a miserable excuse for a human being!* But, like a faithful guard dog trying to prove its worth, Mom jumped in first.

"Yes," she barked, "the pump will break and we'll be left with no water. But of course Teddy isn't concerned about that! What does *he* ever use water for?" Mom tended to have a very poor grasp of mechanical details but was preternaturally provoked by Dad's imperfections.

"Gee, I mean, that piston pump's been there since the thirties," Dad now answered as meekly as he could to avert a quarrel. As much as Dad loved to stir the pot, he avoided direct confrontation at all costs. "I'd like to install a submersible. And to do that, I have to pull out over two hundred feet of steel pipe, which requires that the roof be removed. . . ."

Just as I planned one day to rescue Mom from Rokeby, I felt guilty that I didn't stand up to the family when they ganged up on Dad. When I was alone, I would practice what I would say to them if I had the courage. *I don't see any of you fixing anything. If you don't like his work, why don't you pay professionals to do it? If you don't like his friends, you don't have to speak to them.* But I said nothing. I continually failed as his one and only protector.

AFTER I'D CLEARED the plates and piled them into the warm, soapy water in one side of Grandma's great cast-iron sink, everyone retired to the sofas for more conversation and coffee. Grandma Claire made her usual request of me. "Oh, Alexandra, do play us a little something! I'd love to hear my favorite, 'Long, Long Ago'!"

Reluctantly, I took out my violin. I usually brought it down with me for parties, knowing I'd be asked to play. The folk melody "Long, Long Ago" was from the beginning of volume one of the Suzuki method books, one of the first songs I'd ever learned to play—after "Twinkle, Twinkle Little Star" and "Lightly Row."

A general hush had already filled the room when Aunt Olivia leaned over to Grandma Claire and whispered loudly enough for all to hear, "Will Alexandra be playing us a little 'Twinkle'?" She giggled shrilly, then put her finger to her lips. "Shhh . . . ," she said to herself, and pretended to suppress more giggles.

As I began to play, Grandma Claire sang along under her breath.

Tell me the tales
That to me were so dear . . .

Grandma Claire's melodramatic reaction to the song distracted me. Her eyes teared up, as they always did when I played this, her most frequently requested tune. She sat with her hands clasped, wet eyes expectant, in silent awe.

Long, long ago,
Long, long ago . . .

It irked me that she would cry in the presence of others. What was it, anyway, that she was remembering with so much sadness? Perhaps it was her first love, whom her father prevented her from marrying by sending her off to Florence during her senior year of high school.

. . . But by long absence
Your truth has been tried . . .
Blessed as I was when I sat by your side,
Long, long ago,
Long, long ago . . .

After I played, Grandma Claire and Aunt Olivia settled in to a game of Scrabble. Grandma would play Scrabble any time she could find a willing opponent. She loved to trick her partner into branching out toward the triple word scores in the corners of the board. Once they'd built their words to the edge, she would try to use the high-score letters she'd been hoarding to form her favorite words, like "zit" and "quo." But her performance tonight lacked this kind of strategy.

"Now, Claire, do you want to put down 'medal' or 'meddle'? Are you sure you're all right to play this game?" Aunt Olivia asked.

"Yes, yes—*hiccup*—I'm jus' fine. Now, let me see here. . . ."

Maggie and Diana were in the kitchen washing the dishes, since I had cleared the table. The dirty dishes soaked in one side of the double sink. Diana stood on a stool and sponged off each plate as she removed it from the warm water, then passed it to Maggie, who rinsed it in the other basin and set it on the drying rack.

Dad had fallen asleep in the recliner and was snoring loudly with his head back and his mouth open while Uncle Harry read the *New York Times*.

I too was beginning to feel drowsy from the smell of wine and the soft light from the candles around the room.

I did not wish to wait for Mom and Dad, who were always the last to leave. Maggie and Diana would get driven back up to the big house by Uncle Harry later. So, I decided now was the time to slip out the kitchen door.

So many times, I'd walked home from Grandma's after dark. Tonight there was no moon and it was pitch-black outside. My feet knew the way by heart, knew all the uneven spots along the steep hill.

Peepers from the pond just north of Grandma's house were singing their lungs out in a spring chorus. They sounded like little birds, confusedly chirping at night instead of at dawn. We could never swim in the peepers' pond because it housed water snakes and snapping turtles as well. But it was fun to stand by its edge, listen to the low bass notes of the bullfrogs, and watch the water bugs run on top of the pond's surface.

The spring air was unpredictable. I imagined the blasts of warmth that intermingled with chilly streams to be stray ghosts. My imagination was filled with ghosts.

Nonetheless, I'd learned not to be afraid of the dark. I felt brave, pushing forward into the blackness, listening to the sound of my own footsteps clipping along the hard dirt road.

PART III

ARTISTS
AND DRIFTERS

CHAPTER TEN

THE IRREGULARS

Rokeby was a haven for those who dwelled in the margins. Some of the marginalized refugees included Dad's various charity cases and protégés. Others were legitimate Rokeby tenants.

Almost all our tenants were somewhat bohemian, as the rental houses were not fancy. To Uncle Harry, who had inherited an attitude of scorn for any form of business, the tenants were a necessary evil. Grandma Claire, on the other hand, felt it was her responsibility to foster unity among members of the legitimate Rokeby community—i.e., paying customers—and so she would regularly invite the tenants to dinner.

The old creamery was the center of bohemian life at Rokeby, inhabited by two creative and unconventional women, Debbie and Mimi. Debbie organized seasonal pageants at Rokeby. These would usually be held out in one of Rokeby's fields with fifty or more volunteers—whom Grandma Claire termed "riffraff" and Uncle Harry referred to as "drifters"— awaiting directions as to where and how to move with their papier-mâché masks, banners, and various other props. Larger than life with her wild, frizzy hair; broad, toothy smile; and grand plans and ideas, Debbie reminded me of Glinda, the good witch of the north from *The Wizard of Oz*.

Mimi was a Cuban-American artist with shiny black hair and bushy eyebrows. She had first come to Rokeby as part of a sort of gypsy/vaudeville show with her then husband— a bald, gap-toothed man about twenty years her senior who had started out as her college film professor. They would drive around in a Mercedes, hauling a trailer behind them, and perform sideshows with their German shepherd named Billy Jean, who had been trained to walk on her hind legs in a dress.

While Dad did not identify with Rokeby's bohemians and artists, they all adored him. To them, he was a miraculous mix of classically educated WASP and generous free spirit whose very lifestyle was a masterpiece. As their landlord, Dad was

Rokeby's high priest, who enjoyed the privilege of stopping in at any one of the tenant houses at any time of day or night to eat his rightful share, which was inevitably followed by ice cream with caramel sauce. The tenants would keep this on hand for just such occasions.

Mom, Dad, and I would frequently go over to Debbie and Mimi's creamery for "dinner"—which would involve musical or dramatic performances and storytelling. There, we could always expect to hear exotic music—Gypsy, Balkan, Indian, Jewish—see decorative costumes, and try new recipes. There, one could say or do anything, and it would seem brilliant and entertaining. The creamery was a world of fantasy and pageantry, a world where everyone wore a mask and costume and nothing was ordinary or mundane. And in contrast to the big house, it was a place without judgment, a place of free expression.

Mom would become a whole other person at the old Rokeby creamery, a person I never saw when I was alone with her. She would laugh, recite extemporaneous poetry, and forget to complain about Dad.

Among the other tenants was Alex, long haired, bearded, and lanky, with a large protruding Adam's apple, who rented the milk house. He kept milk crates full of books and records, mostly the Grateful Dead and the Rolling Stones.

There was Matt, who lived in the apartment on the second floor of the coach house, a graduate of MIT's School of Architecture. On summer weekends, Matt would blast hits from musicals like *Oklahoma* and *West Side Story* from his balcony.

Dad promised one day to build the house Matt had designed for his thesis—what Dad called the "house of the future"—at Rokeby.

It was the denizens of this world whom Mom had invited over to the big house for her summer solstice party.

A motley crew of guests wearing painted foam fish on their heads was gathered in the sparsely furnished formal dining room, now hazy with cigarette smoke.

Debbie and Mimi entered the room dressed in their dancing-bear costumes—brown fake-fur suits with stuffed bellies, giant papier-mâché heads, and cute toothy smiles.

Dressed in a suit and not acknowledging anyone, Uncle Harry stood in a corner of the dining room with his back to the guests. These were not the type of people Uncle Harry wanted overrunning and defining the place. His friends were respectable: old roommates from boarding school, fellow club members from Harvard, and various local people interested in historic preservation.

Uncle Harry's obvious disapproval made me feel that by participating in this wild party, I was somehow guilty of doing something corrupt and inappropriate. As he glanced proprietarily at the portraits, I doubted that the ancestors would have been any more approving than he was of the rabble gathered here, in this, *their* house, marring its elegance and embarrassing their staid dignity.

At one end of the dining room were portraits of Great-Grandma Margaret and Great-Grandpa Richard Aldrich—

proud and pompous. At the other end was their daughter, Aunt Maddie, divorced and exiled, as well as Emily Astor, who had been charmed into marriage by the infinitely charismatic Sam Ward. Also present: General Armstrong, William B. Astor, and Grandma Claire. Grandma was a different person in the portrait, with her curls still black; her back still straight; beads of jade around her long, modestly exposed neck; her mouth closed with teeth already clenched, stiffly self-conscious—so beautiful, yet never carefree, not even then.

Grandma Claire never came to our parties. She was only comfortable as a hostess of her own gatherings—necessarily structured around meals.

Some guests lounged in the home parlor—a term we used because "family parlor" sounded too bourgeois—splayed over the room's silk-upholstered chairs and corduroy sofa. Others stood around the gramophone, which had been a personal gift from Thomas Edison to Great-Grandpa Richard Aldrich. As Dad cranked it up, the quarter-inch-thick record began to speed up until it sounded like "Ragtime."

Then interest shifted to the collection of iceboat photographs hanging on the home parlor's wall.

"My father was an iceboater," Dad explained. "FDR gave him an iceboat called the *Jack Frost*. FDR would frequently come visit the Delanos next door, who were his very close relatives, and then stop by at Rokeby. I don't know if he ever knew that my grandmother didn't vote for him in the '32 election. . . ."

Uncle Harry now approached me. "If you see anyone who doesn't belong in other parts of the house, be sure to drive them out."

Uncle Harry wanted to make the younger generation his accomplices in checking up on the guests. In training to drive these undesirables out of Rokeby, we were instructed to interrogate any strangers we saw on the property. Inevitably, however, they would say they were "friends of Ted's," and we would simply have to move on.

I nodded my consent to Uncle Harry's request and proceeded across the front hall, where I came upon an unfamiliar couple drifting around in the drawing room.

The man pointed out the giant, gilt-framed mirror behind the Steinways. The mirror itself was so aged that it was gray and no longer reflective.

"I think that mirror has seen so much vanity that it decided to close for further business. If someone tried to search for himself in it, he'd only be swallowed up."

"A mirror that consumes images . . . ," the woman said. "I'll bet that's what this place will do to a person after a while."

WHEN THE CALL for performances began, I was asked to play something on the violin. I felt self-conscious about asking the guests to stand still and listen quietly and respectfully as I performed the first movement of a Handel sonata. But everyone clapped politely when I was done.

Next, two of Dad's friends, both named Bob, got up to sing some World War I songs, as they always did at our parties.

> It's a long way to Tipperary,
> To the sweetest girl I know!
> Good-bye, Piccadilly . . .

The final act tonight was Mimi doing her famously hilarious Louis Armstrong impersonation, which she would agree to do only once she was good and drunk. She puffed up her cheeks and puckered her lips, and her naturally deep voice grew much deeper as she began to sing "Blueberry Hill."

She held up an imaginary trumpet. "Blow that horn, Louis!" the spectators shouted, cheering.

❋

Throughout the generations, the family's greatest challenge had been coexisting amicably and equably at Rokeby. With co-owners living under one roof, there was bound to be a struggle for dominance. At Rokeby, the nonconformists had always lost the contest.

In the past, Aldriches and Astors had banished family members under the pretexts of bohemianism and insanity. In fact, artists had always been viewed with suspicion at Rokeby. John Jay Chapman perhaps expressed the family's distrust of artists best when he wrote "[Artists] are invariably corrupt and irregular in their private lives and in their ideas. . . . Most of them are foreigners—Italians, Frenchmen and Russians and you know what foreigners are. They don't speak the truth, or pay their bills, or keep the Sabbath."

Mom was an artist. But while she was foreign, she was not at all irregular. When she wasn't chasing after Dad to fix things or give her money, she sat at her drafting table for hours, drawing, etching, painting. As a child, Mom—like me—had been a high achiever. She'd always been the very best student in her class—diligent, precise, and serious. She'd filled albums with incredible childhood drawings. When I was little—before her sadness and powerlessness set in like a disease—Mom had had lots of energy for self-improvement. She'd taken swimming and driving lessons, and had learned English by devouring Agatha Christie novels.

Like the old creamery, our apartment was a bastion of bohemian living on the estate. Only there did Mom freely display her tapestries, engravings, and paintings, as well as her posters from Bread and Puppet—a folkish puppet theater with politically leftist leanings. These latter were rough paintings of some of the central characters of the idealized Communist society: washerwomen in kerchiefs, with oversize papier-mâché heads, and garbagemen in gray workers' uniforms.

"I thought you wanted to get away from the Communists," Dad would chide.

I did not choose to view myself as bohemian. I understood that in my own family, being an unconventional person with artistic leanings was grounds for exclusion, a reason to be deprived of everything, including the right to identify as an Aldrich.

Surrounded by whimsical, unstructured people who did what they pleased whenever they pleased, I genuinely idealized a respectable and disciplined life. I longed to live in a house with modern heating and plumbing. I dreamed of having a presentable car and parents with jobs.

But what could I have been, really, if not bohemian? I was a free spirit who watched unrated, arty foreign films with Mom and dressed in vintage clothing from the local thrift shop. Our living quarters were furnished with random, broken hand-me-downs. We didn't live by rules that coincided in any way with those of the outside world. We never had a dinnertime or matching dinnerware. I had no set bedtime. I did not even own a pair of pajamas.

THE MOST INFAMOUS bohemian to be exiled from the family was Sam Ward, brother of Julia Ward Howe. "Uncle Sam" had married Emily, one of W. B. Astor's daughters. He had been a Renaissance man, infinitely charming and social, loose with money, open-minded, and always looking for a good time. Emily died when their daughter, Maddie, was just two years old. As soon as Uncle Sam remarried—to a Creole woman—the W. B. Astors took custody of their granddaughter and and banned Uncle Sam from Rokeby.

While the vivacious Uncle Sam had clashed with his cheerless, conservative Astor in-laws, he had been extremely popular with most everyone else. According to a *Vanity Fair* article published January 10, 1880 (as part of a series entitled "Men of the Day"):

> Every traveler to the United States, whose lot has fallen to pleasant places, is sure to have met with Sam Ward . . . uncle of the human race. He is the one man who knows everybody worth knowing, who has been everywhere worth going to, and has seen everything worth stepping aside to see . . . a sound scholar, a thoughtful reader, a man of much experience, observation and wisdom, he is yet seen at his best in some act of gentle ministration to the poor and afflicted. His fund of anecdote is inexhaustible. His very presence in a room is enough to put everyone else in good humor. . . . His wit is ready, and his good nature is so great that most Englishmen who have seen New York bring back from it, as one of the most pleasant of their reminiscences, their memory of Uncle Sam.

It was not difficult to see why such a man would not have gotten along with the parsimonious W. B. Astor, the slumlord of lower Manhattan who had left hardly any of his massive fortune to charity. But despite his banishment, Uncle Sam's free-spirited genes seem to have been passed down through the generations.

In the Astor orphans' generation, the line between bohemian and straitlaced had become blurred, as all of the orphans had been unconventional to varying degrees. Only Great-Grandma Margaret had stood apart, a guard of order who kept alive the family tradition of banishing family members at will.

Uncle Archie Chanler was ostracized on the grounds of insanity. He had been left extra money by his mother because, as the eldest Astor orphan, he was expected to take care of Rokeby. She also hoped he would settle at Rokeby and keep the Chanler name alive in the Hudson Valley. But he moved far away, settling in Virginia.

Archie was admittedly a bit odd. He reportedly drove a car that contained a kitchen and a bathtub. When he visited Rokeby, he was said to have eaten grass on the Astor china in the dining room and climbed in and out of his second-floor bedroom window in the middle of the night by way of a ladder, then slept all day. He also experimented with psychic phenomena. For instance, he attempted to channel the spirit of Napoléon Bonaparte. He believed that his subconscious, which he called his "x factor," would send him messages, which he could express through "automatic writing" while in a trancelike state.

But Archie's alleged lunacy was likely a fabrication. His troubles began when he invested money, together with his brother Wintie and Wintie's friend Stanford White, in a few enterprises in Virginia. Archie soon asked Wintie to step down as chairman of the board of the company.

To prevent being deposed, Wintie conspired with their brother Lewis and Stanford White to have Archie declared insane and incarcerated in an asylum. This move put a temporary stay on Archie's power of attorney over Rokeby's finances and cut him off from access to his own money and property.

For four years, Archie stayed at Bloomingdale Insane Asylum in White Plains, New York, under round-the-clock supervision. When Archie appealed to his other siblings for help, they all deferred to Wintie's and Lewis's judgment in the matter.

When Archie finally managed to escape from Bloomingdale, he left the following characteristically witty note for the asylum's medical superintendent.

> *My dear Doctor:*
> *You have always said that I am insane. You have always said that I believe I am the reincarnation of Napoleon Bonaparte. As a learned and sincere man, you, therefore, will not be surprised that I take French leave.*
> *Yours, with regret that we must part,*
> *J.A. Chanler*

Feeling his family's betrayal keenly, this Chanler, who was meant to carry on the family name, took his revenge by changing his surname to Chaloner.

Was Uncle Archie more mentally ill than Great-Grandma

Margaret's brother-in-law John Jay Chapman, who had to be kept in Rokeby's billiard room on the third floor for eighteen months due to delusions and imagined paralysis, and who purposely burned his hand so severely that it had to be amputated? John Jay Chapman was never institutionalized for mental illness. The primary difference between the two men seemed to lie in their relationship to the Astor fortune.

In our generation, the law entitled Dad to the same rights as his brother, so his nuclear family could not legally be sequestered and denied privileges. Yet, in reality, we too had been banished—to the third floor. We lived in the eaves.

CHAPTER ELEVEN

A SEED IS PLANTED

Fortunately for me, the family did not associate classical music with bohemianism. My great-grandparents Aldrich had made a life of music respectable.

It was the middle of June. Giselle was now at Rokeby every day. She and Dad were taking me to my end-of-the-year violin recital. Dad rarely attended my various recitals or school events, because Rokeby's needs always came first. Today, I was too excited that he was finally coming to one of my recitals to worry about why Giselle was joining us instead of Mom.

When I'd asked Mom if she planned to come, she'd snorted in bored contempt. "You know I don't like the sound of children playing the violin. I don't like being around kids at all, if I can help it."

"But I'm a kid."

"Yes, but you're different. You're my daughter."

Mom would occasionally try to disguise her indifference to my activities by telling me how much she valued my independence and my freedom to make my own decisions. She would explain how she had promised herself that she would not be an overbearing parent as her own mother had been. But even then I knew that it was simply easier for her if Grandma Claire took care of me, or if I took care of myself.

Dad, on the other hand, was pleased that I was so self-motivated and musically talented. He doled out compliments about my playing freely as long as he didn't have to pay for the lessons.

Giselle seemed to have permanently claimed the middle seat between Dad and me in the cab of the Chevy. As we started down the carriage drive and Dad shifted the gears, his hand knocked against Giselle's fleshy shin. With each jolt of the truck's shifting gears, the nose of my violin case stabbed her in the arm.

I was wearing my special red concert dress. It had two layers, both red, with satin underneath and stiffened gossamer on top. The top layer was covered with white polka dots that felt bumpy where they'd been painted on. The way it was tight at the waist, then blossomed out into a full skirt, made me think of flying.

But just as I was picturing myself impressing the parents with my performance of the Vivaldi A-minor concerto, we heard a telltale flapping sound.

"Yup, I was afraid of this!" The gears began to grind as Dad hastily shifted down. He swung out to look. "She's flat."

I felt the familiar symptoms of panic—it was as though my heavy heart was leaning against my lungs and making me short of breath. Grandma Claire wouldn't let Dad use her car if Giselle was with us, and Dad would never tell Giselle that she couldn't come. So I'd have to get Grandma Claire to drive me to my recital. I hadn't even told her about the recital because she'd been hitting the bottle so hard lately. While I was ashamed of her red nose and cloudy eyes, at least if Grandma Claire drove me, Giselle wouldn't be there. I didn't want the extra worry of having to explain Giselle to people, especially since no one had really explained her to me. And I didn't want the shame of Giselle's being at my recital while Mom wasn't.

As the passenger door wouldn't open from the inside, I stuck my arm out the window—which was always open because the crank had fallen off—to reach the outside handle. I opened the door and started race-walking down the hill to Grandma Claire's house, my violin case rocking like a buoy in choppy water. The dusty driveway quickly powdered my shiny black patent leather concert shoes, whose slight heels made me feel grown-up.

"Grandma?" My shoes clicked into her musty, salmon-tiled atrium.

"Yes . . . ," a faint voice called in alarm from her room. I found her seated on her bed, slightly slumped over, with her feet on the floor. I didn't want to look at her. Whenever she was "in a state," as Aunt Olivia called it, I had a strong impulse to sit down with her and write up a schedule and a set of rules—a plan for us to follow.

But there was no time for that now. She was dressed in her blue-and-white-checked pants and red polyester blouse—J. C. Penney specials. Her thick horn-rimmed glasses were sliding off her reddened nose. The skin of her neck hung loose like a turkey's. Her eyes looked out at everything at once, overwhelmed as if by something swarming around her.

"Do you need help getting up?" I asked, though it made me queasy to see her, of all people, looking so lost. But I had to suppress my fear, as well as the surge of rage that made me want to beat her sober. I could not fall apart before the recital. We had fewer than fifteen minutes to get there.

"No, no. Just go watch some TV and I'll be right out," she said, not registering that I was dressed up and holding a violin case.

"Well actually, I came to ask if you could take me to my recital."

She stared blankly. "Oh, dear." Her voice was hollow. She

stumbled a few steps as she heaved herself up, as if the alcohol had tilted her brain.

"I'll just wait in the hall," I said. From there I could hear her hiccup loudly, so I hummed the opening to the Vivaldi concerto to block out any more of her sounds.

As she made her way into the hallway, I noticed how her bony hands—crooked with arthritis—were shaking. Slouched way over, she leaned her spidery fingers, white with pressure, against the soot-darkened wall.

Grandma Claire's dry cough reeked of mouthwash. I knew that when she was like this, her driving was life-threatening.

Nevertheless, I preferred Grandma's drunk driving to Mom's expressions of maternal indifference. So I rushed Grandma Claire to the Plymouth. As she lowered herself into the front seat, she left her storklike legs and knobby arthritic knees outside. Then she pulled her legs in one by one—they were too long to fit comfortably—and squeezed them under the steering wheel.

"There, now. *All* ready." She pulled down the sun flap to look in the mirror, put on her red lipstick, then pursed her lips with a smacking sound. She had smeared some over the top and bottom of her lip line and hadn't quite reached the edges of her mouth. But I guessed it made her feel presentable.

Since we took the farm road, we avoided Dad and Giselle, who might still have been with the Chevy on the front drive.

Once we were on the public road, with only minutes to spare, Grandma Claire drove at a snail's pace and braked as she approached each driveway along the way. As she weaved back and forth over the divider, I focused all my psychic energy on the gray pavement moving toward us, and my right hand grew white from squeezing the door handle. *Just stay right. Just stay right. Just stop at the stop sign.* Fortunately it was less than two miles to Mrs. Simmons's house, where both her piano students and Mrs. Gunning's violin students would be performing.

By the time we arrived at Mrs. Simmons's house, the violin students were sitting on the floor next to the glassed-in green room for plants and the piano students sat on the sofas against the wall, all ready to perform.

There was a balcony overlooking the open living room below, which had a black baby grand in its center. Grandma Claire stumbled up the stairs to join the parents seated on the balcony. They were peering over the railing, each one eager for his or her own child to perform. I attempted to ignore Grandma as she gesticulated, pointing out a spot on the floor where she thought I might take a seat.

"And now, we will hear . . . Alexandra," Mrs. Simmons announced in her wispy, whispery voice. She clasped her hands together and smiled in my direction as I rose clumsily, trying not to step on someone's violin or bow.

Mrs. Gunning, my violin teacher, who doubled as an accompanist, was seated at the piano. She was short and slightly stooped yet nimble. She waited for me to set my sheet music on the stand, then played an A so I could tune.

Five years before, on a golden summer afternoon, Mrs. Gunning had ridden her horse across Cousin Chanler's estate and through the fields to Rokeby to ask if I would be interested in taking violin lessons. She had three students for her new Suzuki group and could use a fourth.

Now, I lifted the violin to my chin and nodded for Mrs. Gunning to begin the introduction to the first movement of the Vivaldi concerto. Within moments, I came in with a proud, unapologetic sound. I played with the passion of proving myself, each note clear and bright. I could not play softly, for fear of losing my audience's attention.

As I closed my eyes and imagined Mom sitting up in the balcony, listening with bated breath to my every note, I was suddenly distracted by an awareness of how people must see me. I felt exposed, as if all these people knew about Dad's crazy ways, Mom's depression, and Grandma's drinking. Were they all feeling sorry for me? Were they judging me? I could feel their heavy stares weighing on my hands, my arms, my back.

And for the first time during a performance, I lost my self-confidence. My teeth clenched, my left hand balled up, my chin rest jabbed into my neck, my bow hand trembled. My sound became squeezed, my bowing unsteady, my fingers slippery with cold sweat. In that moment, the whole terrible shame of my life at Rokeby seemed on display, and I began to worry that I would drop my violin.

After the recital, over refreshments, I could hear

Grandma Claire—who obviously hadn't noticed the change in my tone and confidence during my performance—proudly responding to people's comments and questions. She could pass for sober now.

"Well, yes, I *am* very proud of my granddaughter. She certainly doesn't get it from me. . . . Oh, why I know she's talented! But what is it they say about genius? Ninety percent perspiration? My granddaughter is entirely self-motivated, you know. No one ever has to tell her to practice!

"You know, her great-grandfather, Richard Aldrich, was the music critic for the *New York Times* from 1902 to 1923. He was the closest musical relative."

Everything proceeded as it always did. But a seed of something ugly had been planted.

PART IV

ALL IN A
SUMMER'S PLUNDER

CHAPTER TWELVE

A PARALLEL UNIVERSE

With the arrival of summer came the fun that was so lacking the rest of the year. It was the only season when the big house felt lived in, as cousins and family friends would come to stay.

Rokeby in the summertime was the ideal place for children, with its expansive lawns and many dusty and uninhabited rooms to poke around in. We'd have picnics by the river at Astor Point and take afternoon swims at our cousin's private pool.

"Mo-om, can we go swimming now?" Maggie and Diana begged Aunt Olivia.

"Ugh!" Aunt Olivia had learned to moan and groan like Grandma Claire. "Oh, *awll*-right. Just go wait in the car."

"Yeah!" We all cheered, then poured like a flood out of the house, down the back steps, and into Aunt Olivia's black Jeep Wagoneer.

We rolled down all the windows, placed our towels on the hot vinyl seats to protect the backs of our thighs from burning, and left the doors open.

"Mo-om! Can you come *now*?" Maggie shouted up to the kitchen window.

"Awl-*right*! Just wait a minute, will you? . . . Damn children," we heard her mutter.

We waited another fifteen minutes. I told Maggie to honk the horn. She honked.

"Awl-*right*!" Like a peeved hornet, Aunt Olivia shouted out the kitchen window. "If you're going to be so rude, we won't go!"

None of us believed her. We sat there for another five minutes, sweating and fanning ourselves with our hands.

"I'm going to die of heatstroke, damn it!"

"When are we going?"

I told Maggie to call her mother again. She called up.

Then Ben, Aunt Olivia's teenage son from her first marriage, who was just home from boarding school, ran down the back stairs. He was tall, wiry, and shirtless. His chest was concave, as if a meteor had landed there, or as if he were slowly imploding. Ben was the only boy at Rokeby.

"Oh, no! Ben's coming!"

"Ben, you can only come along if you don't play shark attack!" Maggie ordered.

"I'm scared of shark attack," Diana whimpered.

Ben forced us to shove over in the backseat.

"It's too hot!"

"I feel sticky!"

Ben drilled Diana in the arm with his knuckle. She screamed, then cried.

Finally Aunt Olivia came, huffing and moaning, down the stairs. She had on a one-piece bathing suit with skinny shoulder straps, a frilly skirt printed with a jungle of vivid flowers, and a sun visor. She was carrying a canvas bag on her shoulder.

"Okay. Is everybody ready?"

"We've been ready for an hour!" Maggie rolled her eyes.

"Don't you be fresh with me, young lady! Why's Diana crying?"

"Ben punched her for no reason at all!" Maggie tattled.

"Ugh . . . ," Aunt Olivia moaned as she heaved herself into the driver's seat.

Once we reached the neighboring estate, Aunt Olivia let us ride the rest of the way on the open tailgate.

This had been the estate of Great-Grandma Margaret's sister Aunt Elizabeth, who had married John Jay Chapman. Their son, Chanler Chapman, now lived on the estate.

Whenever the question of mental illness in the family arose it was the Chapmans who came to mind. John Jay Chapman was subject to frequent bouts of both serious depression and irrational, impulsive behavior.

Chanler Chapman was also known to have episodes. Once he was arrested and placed in an asylum for shooting up a local tavern. And on the morning after his mother's death, he reportedly came over to Rokeby with a shotgun to see his aunt Margaret. He found Great-Grandma Margaret having breakfast in bed, and, without explanation, aimed the gun right at her. At this, she nonchalantly continued chewing; smiled her squinty, condescending smile; and calmly pushed aside the barrel of his gun.

COUSIN CHANLER CHAPMAN'S pool was a rectangle of blue in a lonely, prickly brown field, and he sat beside it in a portable lawn chair, Jamaican home attendant at his side. He was an eighty-year-old Scottish terrier of a man, with fuzzy salt-and-pepper hair and bristly whiskers. A Polaroid camera hung on a string against his sparse gray chest hair, as he was always hoping to catch some shots for his private gazette, the *Barrytown Explorer.*

Cousin Chanler had started the *Barrytown Explorer,* a newspaper substantially about himself, in 1948. It mainly chronicled his thoughts and deeds but occasionally featured articles about people in the local community. As Cousin Chanler's newspaper was paid for exclusively with his own private funds, he could afford to write whatever he wished. And this had been his general philosophy of life: to say, write, and do anything he damn well pleased.

Like Dad, Cousin Chanler had created his own universe. Yet as he had been the sole heir of Aunt Elizabeth's estate, he had no vigilant co-owners to thwart his autonomy.

Our posse now approached like a wolf pack in our beautiful bathing suits—mine was white with green leaves interspersed with stunningly red ladybugs—our towels over our shoulders, threatening to create a ruckus of splashing and screaming.

Since he'd become old and infirm, Cousin Chanler had been living in the red ranch house on his estate, while his mother's mansion remained uninhabited. Only once, before he moved out of it, had I seen the mansion's four square rooms on the ground floor, with their handsome French doors; twenty-foot-high ceilings; gigantic, threadbare Oriental rugs; and ornate marble mantelpieces. While it was small in comparison to Rokeby, it was cozier.

Cousin Chanler now started to pull himself up on his walker to properly greet our posse. We were frequently reminded to be polite or we wouldn't be allowed to come swim there anymore. During our patient ritual of shaking hands and repeating our names, he grunted approval with a wet streak of drool in the crack of each corner of his mouth. By the time we had thrown down our towels and were running for the water like animals fleeing a forest fire, he had the boxy camera pressed against his focused journalist's eye.

Within minutes, we were all screaming in terror of the "shark" in the pool. Ben's idea of fun was to swim under us, grab us by the legs, and pull us down. We screamed, swallowed the chlorinated water, and struggled to the surface to catch our breath, panicked as much by our unfazed chaperones as by Ben's strong hands.

Since Ben's teasing ruined my plan to swim laps, I climbed out to practice my back dives off the diving board instead.

I got onto the board and stood with my arched back to the pool, gripping the edge of the diving board with my toes and letting my heels hang over. I held my arms up straight with my fingers together and bounced slightly. Then, like melting wax, my head and back collapsed backward. My hands entered the water first, without a splash. They were followed by my arms, head, torso, legs, and finally, my perfectly pointed toes, like a dagger silently entering into flesh.

"Fine form, whatever your name is!" Cousin Chanler shouted at me when I resurfaced. "Whose daughter are you?"

"Teddy's," I answered from the water. I knew to be wary of this grumpy old man.

"That old son of a bitch Teddy? Has a pretty daughter, does he? Do you read Shakespeare, Teddy's daughter?"

Saul Bellow had once rented the apartment in Cousin Chanler's coach house. Bellow reportedly had been so affected by this cranky character that he'd modeled Eugene Henderson in *Henderson the Rain King* after him.

"No, not really," I answered, slightly ashamed of myself.

"What do you mean, no? Didn't Aunt Margaret ever read Shakespeare to you?"

"I never knew my great-grandmother."

"No, no. Course you didn't, course not! Damn foolish question! A shame Ophelia couldn't swim as well as you!"

A few years earlier, Cousin Chanler had paid Dad to go to Florence and rescue Sty, his son, from the flea-ridden storefront where he'd been slowly dying of alcoholism and malnutrition amidst piles of newspapers and dog excrement.

"Once Sty had sat down in the plane, he got into the hooch and went absolutely haywire." Dad loved to repeat the story. "He upset a cart of drinks and started screaming obscenities at one of the stewardesses and throwing bottles around inside the plane. The stewards got all the other passengers out of our part of the plane and approached us with CO_2 fire extinguishers.

The plan was to freeze us with the CO_2, grab us, tie us up, throw us into the bathrooms, lock the doors, and head for the nearest airport. I had to placate the captain with the lie that this man was my patient, and he would die if he were treated that way."

When Sty first arrived back in the States, he was unsteady on his feet and shuffled obsequiously through our apartment door, his long, spidery fingers resting on his bloated abdomen. His frizzy grayish-brown hair had grown into an afro, and he wore a simpering, vacant grin. Ever since then, Sty would call us—the tolerant relatives who were his only connection with the outside world—on a daily basis, drunk and in need of someone to vent to about his "son-of-a-bitch father."

After an hour of swimming, we all piled into the backseat of the Jeep, where we sat on our damp towels.

"I s'pose *I* am feeding everybody today," Aunt Olivia mumbled to herself. "Claire's been sucking on the bottle lately and is in no condition. . . ." The expression sounded so crass: "sucking on the bottle." Grandma Claire wasn't some drunk passed out in a gutter.

CHAPTER THIRTEEN

AN IMPORTED ORDER

In early summer, Grandma Claire would start frantically preparing the house for her daughter's annual visit from Paris.
We would begin our cleaning with the formal guest rooms on
the second-floor landing.

"Ugh! Oh damn it all! This, ugh, damn mop!" Grandma
Claire had fallen into her usual frenzy. In Paris, Aunt Liz and
her husband, Uncle Alessio, lived in style, with a Mercedes

sports car and au pairs for their daughters. Grandma's fear of comparison between their comfortable accommodations in Paris and the tattered state of things back home made her panic. She needed things to look as they once had, but the pressure to transform Rokeby overnight was overwhelming.

"Ugh, dear me!" Grandma now moaned, her buckets overflowing with cleaning fluids and her mop. As she heaved her weight up each dirty step leading to the second-floor landing, the mop's stick would knock into her fluffy hair. Two new pillows, still wrapped in plastic, were under her arm, and a fan hung off her fingers. "Here, Alexandra. Can you relieve me of some of these?" Her glasses kept slipping down her sweaty nose, and their lenses were steamed up. "Oh, these damned steps! How can anyone *live* in this god-forsaken house?"

As she spoke, Grandma didn't look at me once. Clearly her thoughts were not on the present but on a dreaded possible future: *I am a source of disappointment and shame to my daughter. Liz will be displeased. I must transform this house.*

Aunt Liz was the only one from the younger generations who had really gotten away. She had married an Italian architect who wore ironed slacks, polo shirts, loafers, and Swiss watches, and smelled of cologne. Whenever Uncle Alessio walked, loose change would jingle in his front pockets, while the wallet in his back pocket would form a square outline. None of the other men I knew ever carried change and a wallet, wore a watch, or smelled of cologne.

They had two daughters. Anna was six and Judy was a toddler.

Grandma and I dispelled the stillness of the front rooms by opening all the windows and setting fans whirring on the sills. We made the beds with sheets newly purchased from Caldor and took Great-Grandma Margaret's moth-eaten bedspreads out of the linen closet. We put soap and toilet paper in the summer bathrooms. Then I swept the white stairs and front hall and oiled the parquet. I enjoyed watching the oil transform the dusty, worn wood into something shiny and new.

When summer began and the third floor became unbearably hot, Dad and I would move into our "summer rooms"— which were among the formal front guest rooms off the second-floor landing. Mom chose not to move down to the second floor for the summer, perhaps savoring her privacy.

Each of the four front guest rooms had a name. My summer room was known as the "rosewood room." Where the building had settled, there were rips in its beige wallpaper. The paper was covered with the motif of a pheasant whose small head and condescending eyes faced a highly dignified woodpecker on another branch. The woodpecker's crest of feathers fluffed around its neck like the fashionable collar of a proper nineteenth-century gentleman.

Dad's summer room was still known as "T'Amey's room," although Aunt Amey Aldrich—Great-Grandma Margaret's sister-in-law—had long been dead. The room always looked as though it had been ransacked, with Dad's clothes, books, and newspapers strewn about the bed and floor.

We called a third room "Grandma Aldrich's room," as it was where Great-Grandma Margaret once slept. With a stubbornness that matched her spirit, the room had remained unchanged since her day. The bed, the table at the foot of the bed, and the large framed mirror were all matching pieces of the same set: light chestnut, trimmed in a darker brown. I was sure I never wanted to recline in this bed and see my lounging reflection in the tall standing mirror directly opposite. The floorboards, partially covered by a moth-eaten green rug, were faded from the sunlight that constantly streamed in from the east. As there were no objects on any of the room's surfaces, aside from two table lamps, it was clear that no one lived in this hollow ghost of a room. I often imagined my great-grandmother in her old age, sitting upright against the high headboard of her bed, white with age in her cotton nightdress, her tight jaw snapping like an alligator's as she ate her breakfast.

DAD'S LATEST BOARDER, Walter, had been given notice that the guest rooms would be off-limits during the summer months. After his alleged grave-robbing career in Mexico, Walter purchased an abandoned and decrepit mansion down the road from Rokeby, perhaps in the hopes of one day joining the ranks of the Hudson River aristocracy. He was to live with us while he worked on renovating this new house of his, but it seemed to me that he was always at Rokeby and never working on his renovations. What Dad found most entertaining about Walter was that his history was opaque. Dad liked to joke that Walter had several pseudonyms because he was running from the law.

Walter was one of the few men whom Grandma Claire did not see as a potential danger to her granddaughters. "He's one of those, you know . . . *homos!*" she tried to explain to us.

Walter had never said he was gay. People assumed he was because he spoke in a mock British accent and brushed back his naturally greasy hair so that it appeared to have been gelled, with a bump waving out to each side. He would slide like a shadow through the halls and stairwells, descending the stairs or slipping through doorways silently.

We knew about "homos" from one of Grandma's stories, about the "Barrytown Boys," who were a gay couple who lived down the road. They were, in Grandma's words, "perfectly nice, even if they were—you know . . . *homos* . . . ," so she had invited them to a long-ago Rokeby picnic. Everyone was out on the western lawn when Grandma suddenly noticed activity on the roof of the big house. "It was little Teddy leading those homos around on a tour of the roof!" At that point, Grandma almost fainted with fright at what they might have done with her son up in the tower.

But now it was the 1980s, and although there was no need to worry about Walter molesting us, she perceived another threat.

In Grandma Claire's defense, this was when HIV was still first being discovered and very little was known about how it was spread. During our big cleanup that year, Grandma pinned a sign on the guest room where Walter had been staying for the past few months. It read: KEEP OUT. AIDS!

She then proceeded to spray all the rooms and bathrooms with Lysol.

❧

Aunt Liz brought with her the kind of rules and structure I craved. Her children had bedtimes, bath times, and mealtimes—during which she and Uncle Alessio enforced table manners: no elbows on the table, no talking with your mouth full, no using your fingers to pick apart food.

Though Aunt Liz's family shared our kitchen, I remained on the margins of this temporary, imported order. Their food would be stored neatly on one untouchable shelf of the fridge, in the midst of our scant supplies. At mealtimes, Aunt Liz would parcel out very precise portions: half a hot dog, half a slice of bread, a glass of orange Tang or Ovaltine, two baby carrots. I would go to the fridge during my cousins' feeding time and grab a cold leftover chicken leg or a piece of Wonder bread and sit on the periphery of their ordered meal.

Dad never brought home TV dinners in the summertime because Aunt Liz didn't approve of them. This was probably because their name implied that one should watch TV while eating them. To Aunt Liz, anyone who watched TV—or chewed gum—was some sort of degenerate, and she would let them know it. "Chewing gum is for the lower classes and watching television rots the brain!"

Needless to say, we did neither during the summer.

CHAPTER FOURTEEN

REPOSSESSED

I made many trips between my winter and summer rooms to carry down my books, music stand and sheet music, summer clothes, bathing suits, toothbrush, record player, and records. Among the recordings I couldn't spend the summer without were David and Igor Oistrakh playing Bach's *Double Violin Concerto,* the Beaux Arts Trio playing Beethoven's *Archduke,* Stern playing the Beethoven and Brahms violin concertos, and Rudolph Serkin playing Beethoven's *Emperor* piano concerto.

Apart from what I'd carried downstairs, nothing in my summer room belonged to me. There was an elaborately engraved writing desk between the windows—from the era of inkwells and feather pens—that I wouldn't dream of touching. Great-Grandma Margaret's linens still occupied a closet to the right of the bed with a full-length mirror in its door and a brass key that was never to be removed from its lock. Between the fireplace and the door was a low, hard chair—on which I would never sit—with a very tall back and springs leaking out the bottom.

OUR COUSINS FROM France had been at Rokeby for about a week when, on one of my trips to my third-floor bedroom— which boiled with heat and wasps—I instantly sensed something was wrong. I took a cursory inventory: all my big dolls were resting comfortably on their sofa. Even the antique Russian soldier was there. But I froze in my tracks when I noticed an empty space next to the sofa. My dollhouse, together with its furniture and inhabitants, was gone. The one thing I identified as mine alone.

This was an act of war!

Panicked like a mother whose child has gone missing, I started searching the third floor. Past the elevator; among the trunks packed with Great-Grandma Margaret's undergarments; through the several storerooms stuffed with broken chairs and unusable, urine-stained horsehair mattresses.

Nothing.

Then down the steps. Past Aunt Liz's room and the winding passageways. Onto the second-floor landing and past Aunt Alida's portrait under the webbed skylight. Into Dad's summer room, where the bed was piled high with newspapers. Into Great-Grandma Margaret's chestnut-brown ghost of a room. Through the dressing room and around to Aunt Liz's back door. *Knock first. Poke your head in.* Nobody and nothing but the pale yellow peeling walls and dusty bookshelves. Then into the little hall at the foot of the third-floor stairs again, and across to the room with the three little beds, where Anna slept.

And there it was. Intact. "Oh, thank goodness!" I said aloud.

Then the heavy heart and shortness of breath. *You've been repossessed. No longer mine.*

"Aunt liz took my dollhouse and put it into Anna's room," I complained to Mom.

She just shrugged her shoulders. "That dollhouse never belonged to you anyway. It's the property of the house." Mom felt she had no choice but to side with "the owners," which to her meant Grandma Claire—who, in turn, never dared oppose her daughter.

How could a house own property? And surely I had a

long-term lease on it. It had been in my room all these years.

"And what about my dolls' furniture?" I asked.

"The furniture goes with the house," Mom informed me.

When I told Grandma Claire about the dollhouse, she said, "You're too old for dolls anyway. And your cousins don't get to be at Rokeby very much."

In any conflict I might have had with my aunts or their children, Grandma Claire would always take their side over mine, and I was always outnumbered.

"But she should have asked me at least," I protested, still seeking validation.

"*She* doesn't have to ask *you* for permission. *You* are a *child*!"

But wasn't the point here that I was no longer a child? "But you just said I'm too old for dolls. . . . Anyway, it means a lot to me." They could take the big house and all its portraits, just leave me my dollhouse! But mine was an army of one. One child, unskilled in the art of domination, untrained in taking back what is rightfully hers.

"Now, *please!*" Grandma Claire pleaded desperately, as if I were trying to force her to commit some unspeakable act.

I would never get my dollhouse back. It would remain in their summer quarters until I'd lost the urge to play with it.

CHAPTER FIFTEEN

INDISPUTABLE

Grandma Claire had sworn during her third pregnancy that if she gave birth to another boy, she'd take it down to the river and drown it. It was fortunate, then, that she gave birth to Aunt Liz.

For a hundred years, Rokeby had been controlled by women. Like Victorian children, men in the family were expected to be seen but not heard.

When Great-Grandma Margaret—at the advanced age of thirty-six—married Richard Aldrich, the *New York Times* music critic from Providence, Rhode Island, she did not move to live in her husband's home. Contrary to convention, Great-Grandpa Richard moved to Rokeby and agreed to keep all the property—both Rokeby and Great-Grandma Margaret's house on Manhattan's West Seventy-fourth Street—in his wife's name.

She, of course, had chosen a spouse who was easy to dominate. Great-Grandpa Richard had a severe stammer and hardly spoke, so Margaret did the talking. She decided on the topics to be discussed. And they lived in her houses.

Great-Grandpa Richard was busy writing musical criticism for the *New York Times,* anyway. The only rules he ever laid down were these: they were never to invite a musician to play or sing at their house in New York before he had written a criticism of their public performance that year, and Margaret could not influence his thinking by discussing a performance with him before he'd reviewed it.

Great-Grandma Margaret never needed to banish her son, as Grandpa Dickie had been quiet, amicable and compliant. He was never a threat to her control of Rokeby.

Likely because Great-Grandma Margaret had found it easier to deal with her son when he was sedated by alcohol, she failed—or refused—to see how sick he really was. Even as Grandpa Dickie's face fell into his dinner plate, Great-Grandma Margaret continued to assert that there existed a strong no-drinking policy at Rokeby, ever since she had taken up the cause of temperance around the time of the First World War.

"If the boys in the trenches can't drink, then neither will we!" she'd say.

Dad, on the other hand, did threaten Grandma Claire's control of Rokeby. Unlike his male predecessors, he would not allow Rokeby's matriarchs to cramp his style.

"GET OUT!" GRANDMA'S voice cracked like a young adolescent boy's, her engine in high gear. She was facing the pump house, where Dad had been replacing the antiquated pump. "Get off my property!"

Several open beer bottles were scattered on the grass, a sure sign of the presence of "riffraff." Grandma picked up a bottle and hurled it. Her movements were stiff and uncoordinated, like an insect's, and she missed the pump house on her first throw.

Dad was on the tractor on the other side of the pump house, pulling out a long pipe from inside the well. The roof had been temporarily removed.

"I know you're in there!" Grandma Claire struggled to make herself heard over the roar of Dad's tractor engine as she threw one bottle after another at the pump house and at Giselle, who was evidently inside.

Dad was gunning the engine of his John Deere to drown out his mother. He backed down the hill, the end of the fifty-foot-long pipe tied to the tractor bucket, like an endless, shiny black esophagus being ripped from the deep throat of the well.

"Go back to France, you harlot!"

Along with their mother, Giselle's children were constant visitors that summer. One day, when all the kids had been playing in Grandma's backyard, I'd followed Patricia—at seven, Giselle's oldest—to the swing set and positioned myself in front of her so she couldn't swing.

I needed some confirmation of the situation between Dad and Giselle. No one had discussed it with me, and I didn't dare broach the subject of sexual impropriety with an adult.

"Patricia?" I asked, impatient for the truth. "Have you ever seen Teddy and your mother kiss?"

Patricia shook her unbrushed head of hair and avoided making eye contact with me. "*Je ne comprends pas.*"

"You know, *kiss*?" I frantically made a kiss in the air. "My *papa* and your *maman*?" I pushed the tips of my index fingers together and twisted them around. "Kissing? On tractor? In barn?"

"*Non!*" The little girl shook her head fervently, still not looking me in the eye. "*Non,*" she protested.

As Patricia tried to get off the swing, I held the two chains. But just then Grandma Claire stepped out into the yard. "What's going on over there? Alexandra, why are you harassing that poor child?" Apparently, the child of Grandma's enemy was not her enemy.

ONE MIDSUMMER'S NIGHT in the rosewood room, I got the confirmation I'd been craving.

I was in bed reading *The Hobbit* and listening to Isaac Stern playing the Beethoven violin concerto. As the kettle drums boomed at the beginning of the first movement, I imagined the goblins riding through the forest, getting ever closer to the hobbits and their friends. When the solo violin entered, spiraling upward, I imagined the little hobbits themselves climbing up their treacherous mountain paths, trying to get to the ring before the goblins found them.

I put my book down, but I couldn't sleep at first. Here in my summer room I was often distracted by the menacing wooden canopy that hung over my bed. Its purchase on the wall was mysterious as it had no visible hooks or cables. The room's two curtainless windows were gaping black holes that I always imagined might conceal someone peeking in from the muggy night.

Once I'd finally managed to doze off, I was awakened by voices. I heard Dad in his summer room next door. And I heard a familiar lilting voice, soft and feminine. I thought I could recognize the jingling of bangles. I heard Giselle's laughter.

My eyes were wide open now.

In Giselle's defense, it was obvious that she had found her soul mate in Dad. Neither was fond of bathing. Both came from money but chose to live and dress like beggars. Both did exactly what they wanted, without any regard for what anyone else had to say about it. And clearly, Giselle needed to be with Dad constantly and was willing to suffer greatly to that end.

I didn't know why Giselle was this way, but Dad's defiance, lack of personal hygiene and low self-esteem were the natural effects of being raised by a pair of alcoholics. This explained why he let himself be the family's whipping boy. This explained his dysfunctional love life. This was at the heart of the mess in both his bedroom and his barnyard.

I feel certain that Dad's first attitude toward Giselle had been his typical one of indifference. He had never been known to tell anyone not to come around; in fact, I doubted if he was even capable of saying the word "no." But when Dad saw the violent reaction Giselle's presence incited, it seemed to have turned into a gratifying rebellion for him. Both of them thrived on being a notorious duo.

I'D OCCASIONALLY CATCH Mom complaining to Dad about Giselle, telling him to send her home where she belonged. But Mom would stop talking about it whenever I appeared. Even Mom had a sense of propriety about this shameful subject.

Mom—usually so confrontational—also appeared to be afraid of saying anything directly to Giselle. Perhaps the insult was too painful for her. Perhaps Mom was more sensitive than she let on.

I wasn't sure if Uncle Harry even knew about Giselle, so preoccupied was he with his work, and so oblivious to the details of Dad's private life. Aunt Liz tried to get Giselle to leave by telling her sternly that she wasn't welcome. But since Dad wouldn't agree to force her out, Giselle seemed destined to become yet another item in Rokeby's large collection that would remain there, unwanted, collecting dust.

CHAPTER SIXTEEN

ANIMAL WARS

Dad kept his Rokeby dream alive, in part, by maintaining a semblance of the former farm. He grew corn and hay, harvested and mowed the fields, and kept farm animals: a pig, a horse, and two goats.

Working within his limited resources, Dad had to be

creative about keeping these animals. His giant pig, Egbert, ate the rancid pie dough discarded by the local pie factory. He lived in the old cement icehouse, which had been used to store ice cut from the frozen river before the advent of electric freezers. Dad almost never let the pig outside to eat. Instead he boarded up the large rectangular entrance to the icehouse with a slab of plywood, which he would slide open and reach around to drop in the pie crust. Whenever I asked Dad if I could get a peek at the pig, which had taken on legendary proportions in my mind, he'd respond, "If the board gets moved over too much, seeing the light might cause him to charge. And he is big!" Whenever I passed the icehouse, I'd imagine the mad, massive pig charging through its barricade, impaling me with its giant tusks.

Dad had rescued his goats from a laboratory. The research they'd been subject to had left them with plastic intestines. Dad kept the goats in the yard of the milk house surrounded by a makeshift chicken-wire fence and fed them, among other things, aluminum foil from our TV dinners. "They'll eat anything. Watch this!"

It made Grandma furious that Dad kept goats when he couldn't feed his own family.

GRANDMA ALSO HATED that Dad kept a horse named Cricket—his favorite pet—and often threatened to euthanize the animal.

She would rationalize: "You don't feed that creature properly. That horse is all bones, half dead already."

To save Cricket from his mother, Dad secretly brought him to live in Cousin Chanler Chapman's barns. As I got older, the horse began to take on a tragic aspect, banished as a proxy for Dad.

Aunt Olivia, on the other hand, kept four horses in Rokeby's stables. She gave horseback riding lessons in the lower field below the big house, where she would stand with her long black whip in the middle of the track created by the horses' hooves and shout instructions to students in the ring around her. She was often dressed in beige riding britches and black riding boots.

Aunt Olivia's horses included a sixteen-year-old bay, a palomino, and two white ponies. The bay had *dignitas*. He was very tall, with a black mane and tail. Sometimes I would see him get into kicking fights with the palomino out in the pasture, and I'd wonder why Cricket couldn't be out there with those horses, rather than penned up alone inside a dark stall all day.

Sometimes, I'd go with Dad down to Cousin Chanler's empty barn to feed the lonely horse. Cricket's rarely brushed coat was usually caked with mud.

"Good boy." Dad would pull a soft dirty carrot out of his pants pocket and hold it out on his flattened palm, while Cricket's fuzzy lips nipped at it, leaving a bit of drool on Dad's palm.

* * *

ONE AUGUST EVENING, Dad entered our kitchen wearing—beneath the usual layer of motor oil, dust, and sweat—an unfamiliar pall. In his hair were stray pieces of hay, and his gray eyes had the cold, open stare of an Arctic wolf.

"What did Grandma do with Cricket?" I knew only what anyone from the family would know: if Cricket was missing, the only explanation was that Grandma had made good on her threat to put the horse down.

Dad bore down on me.

"Where's Cricket?"

He expected me to give him a straight answer. But would he have given me a straight answer if I'd asked him about his relationship with Giselle?

"I'm not sure," I answered.

Dad rushed off. His steady footfall, which usually thundered through the front hall, had been replaced by something faster and lighter.

The next morning I found Grandma Claire standing on the front stoop calling her white Labrador retriever.

"Bi-aan-ca! Yoo-hoo! Here, girl . . ." She shook some dry dog food into a metal bowl. "Yoo-hoo. Bianca!" She whistled. "Oh, good morning, dear! Have you seen Bianca? It's the strangest thing! I don't know where she could have gone."

We both knew Bianca wasn't capable of going far. She

was very overweight because Grandma Claire fed her three times a day.

"Oh, dear me . . ." Grandma moaned.

So Mom and I went searching for Bianca along the dusty farm road in the hot August sun. Nature had grown still, as if holding its breath. The fields were brown with dead grass.

I looked hard, half-expecting to see the fat white Lab drinking from the stream by Grandma Claire's house, a canine ghost amid the dark green foliage.

Then, all of a sudden, a magnificent ball of orange rose over the roofs of the barnyard, as if the sun were rising at the wrong time and in the wrong place, its edges blurring and blending as they expanded outward.

"Holy smokes!" I exclaimed. "What's that?" I'd never seen so much fire. Mom and I both started running toward it. "Maybe it's the yellow house." It seemed an appropriate end for Sonny Day, with his heliophilic name, to be cremated by the sun itself.

"It's too far north to be the creamery."

"Maybe it's the greenhouse." Once all glass, the original Rokeby greenhouse had been converted into a residence by Dad in the late 1960s, with the help of a Bulgarian named Boris, who was best remembered for his dramatic recitations of Poe's "Raven" in Bulgarian.

Mom and I soon found the tenants congregated at the top of the hill.

The lower barn—a solid three-story gambrel-roofed barn—was on fire. Eaten by flames, it was mesmerizing: melting, dying with more fanfare than it had ever known in its lifetime. The lower barn had always been a dark place, its floors lined with loose straw, smelling of death. It had never been alive with animals in my time, only dead with skeletons of cattle in some of the stalls. Abandoned, it felt to me as if it had already been dead for years.

"Save it, damn it!" Uncle Harry suddenly torpedoed through the crowd, sweating in his suit and tie. "I'll catch whoever's responsible for this!" He ran down the hill toward the barn with one hand in the air, his index finger pointing upward like a torch-bearing messenger rushing between villages with urgent news. Then he suddenly disappeared from view.

"He's fallen into Ted's ditch!"

The ditch, which Dad had been digging since spring, was like a moat around a fortress, making the lower barn inaccessible to fire trucks. While it saved Uncle Harry, it doomed the barn.

Good-bye, old barn.

AFTER THE FIRE, Bianca returned to Grandma's, looking thinner. One-armed Roy had found her locked in one of the junked cars out in our woods. The cars, thirty in all, were lined up, hoodless and paintless, with tires strewn about, pools of broken glass next to some.

In the wake of Cricket's disappearance, only Dad had a motive for locking Bianca in a junked car.

"Did you know anything about this?" Grandma asked me.

I was caught in the middle. I had seen revenge in Dad's face when he asked me about Cricket, and I'd sensed that rage had temporarily transformed him, but did that make me an accomplice?

"No!" I said defensively. I felt accused by her knowing look: her lips pursed in annoyance, her eyebrows furrowed, her eyes glaring at a slight angle over the rim of her glasses. She could get a false confession out of me if she glared at me this way for long enough.

But instead, Grandma Claire turned the criticism against herself, because someone had to take the blame. "I must have done something wrong to deserve such a son. . . . Well, I'll be dead soon, anyway. Then you'll all be rid of me."

Grandma Claire would often say this, hoping it would get us to repent and mend our ways.

Despite her implied resignation, Grandma retaliated anyway, by disposing of Dad's goats. For the time being, exactly how and where she disposed of them remained a mystery.

By mid-August, swimming lessons had come to an end. Mrs. Gunning had gone on vacation, so my practice record sat dry and empty. *Month of August: no progress whatsoever.*

It was therefore a great relief when Giselle's husband, Jacques, arrived on our doorstep—just days before our annual Rokeby square dance. Finally an authoritative adult had decided to intervene.

He charged into our kitchen, wearing a dress shirt and tie, presumably on his way home from work. "*Un mot, monsieur, s'il vous plaît,*" he said to my dad. He ignored Mom and me, adding only, "*En privé.*"

Dad led the way, through the windowless pantry, redolent of tuna and chicken cat dinners; past the giant cupboard that stored the green-and-gold-rimmed Astor china; past the portrait of John Jay Chapman, the tapestry of Pompey, and the engraving of George Washington. Jacques was at his heels, a full head shorter, with a square frame, an inflated chest, and hunched, rounded shoulders. With his multiple chins tucked into his neck, he was a bulldog of a man.

All I could hear through the door was a fast stream of incomprehensible words, with an emphasis every fifth word or so, like small gunshots. The conversation was clearly one-sided. I felt humiliation for Dad as he got a severe scolding. But did he feel it? I wondered. Later, Mom reported that after Jacques had given Dad an earful in French, he had slapped Dad hard enough to give him a bloody nose. I imagined Jacques challenging Dad to a duel, to defend his French honor, and

Dad explaining that it would be impossible just now, as none of his weapons were currently in working order. Aside from the collection of rusted swords in cracked, crumbling leather sheaths in the front library, Dad had only various shotguns that he used for hunting.

He used one of these to kill a pig for the square dance.

CHAPTER SEVENTEEN

A MANLY ENDEAVOR

Dad had started up our tradition of an annual square dance in the early seventies. It included a potluck picnic, but Dad provided the meat. Every year, he would buy a small live pig several days in advance.

This year, we kids peered in to see the pig lying down on the backseat of a crashed-up, tire-less 1968 Volvo that had been lying around the barnyard for some time. The animal's grotesquely fat belly heaved as it panted.

"Here, piggy, piggy, pig! Awww."

It stuck its snout toward us and sniffed the air through the crack in the window. The flaps of its snout around the nostrils wiggled visibly. We tapped the glass, laughing.

Dad and company had unloaded the pig into this junked car, the only one in Dad's collection that still had all its doors on. At first, the animal had squealed and torn up what was left of the car seats, but eventually, the heat inside the vehicle had subdued it.

The pig was too lethargic now to get up from its reclining position on its side, with its fat top legs folded politely over its bottom legs, its cloven hooves pointing at a forty-five-degree angle, in a permanent tippy-toe stance.

The pig could suffocate in the closed-up car under August's noonday sun, if someone didn't shoot it soon. But no one seemed to worry about treating the pig humanely, because we all knew it would be killed shortly.

A bit later in the day, Dad climbed into the Volvo's driver's seat with a shotgun. I couldn't watch, so I hid behind the barn. There followed a slightly muffled explosion inside the vehicle as he shot the pig in the head at close range. Once the worst was over, I returned to the scene.

As Dad and a few of his buddies dragged the pig out by its hind feet, it hit the dirt with a thud. Then they threw it into the front loading bucket of Dad's tractor. Dad raised the bucket, now carrying Roy as well, up to a branch of the Chinese chestnut tree that would drop sweet edible chestnuts encased in their prickly shells onto the groundskeeper's lawn.

Standing on the edge of the bucket, Roy—who was extraordinarily adroit for a one-armed man—wrapped a rope several times around the sturdy branch, then tied the other end of the rope around the pig's hind feet.

"Okay. Lower her down!" he shouted. As the bucket descended slowly, the pig's body began to slide out of the bucket, until—*whap!*—it fell into midair with a snap as the rope tensed between the branch and its ankles. The long peach-colored sack of flesh swayed as it hung. The single bullet wound in its head leaked one rivulet of red—a clean shot.

"Okay. Let's slice 'er open!" Dad said.

Roy brought a stepping stool and butcher knife to the pig's side, climbed up the three steps, and placed his stub on the pig's flank to steady it. Its beady blue eyes, with white eyelashes, were open. Its meaty ears hung down. Its snout was still.

Roy then made one silent cut from the throat all the way up, up the gut. The edges started to redden as soon as the blade created them; then the blood burst through the seams, and Roy jumped away. The stepping stool got rained on by the steady red downpour.

Sonny Day now rushed down the porch steps of the barnyard's yellow farmhouse. "Keep that blood off my lawn, you sonsofbitches! Goddamnitall! And stay away from my roses!" he shouted. He stood protectively next to his circular rose garden, immured by chicken wire, and rested his weight on his hedge clippers. His hair was thinning and white. Behind him was a dying crab apple tree, gnarled and hollow.

Dad wore some rubber dishwashing gloves that he'd swiped from Mom. He climbed onto the stool and sank his hand into the pig's belly, up to the elbow, then pulled out endless lengths of the fat, gray, wormlike intestines. He reached in again and pulled out a pink sac. Everything was coated with blood.

We kids stood on the periphery, grimacing and hunching over, exclaiming:

"Yuck!"

"That's the most disgusting thing I've ever seen!"

"*Ewww!*"

The pig was left to drain, with its entrails resting beside it in two large metal pots—also from Mom's supply. Some of the intestines hung over one pot's brim, too long to fit neatly inside.

Aunt Olivia's horses got a few hours' respite when the flies abandoned them to visit this new and delectable supply of free blood.

ON THE MORNING of the square dance, I found Dad and his buddies sitting out on our front lawn on metal folding chairs. Today they were the keepers of the spit, seated around a pig corpse that had been impaled on a pole, with handles for turning it over the fire. Its stomach cavity and mouth were stuffed with apples and wrapped in chicken wire.

About midmorning, the beer kegs arrived, like oversize silver bullets. They came topped by black hand pumps and rubber hoses with taps on the ends for spraying the foamy, golden ale. At the party, the kids would inevitably sneak a little, because it was fun to squirt the bitter, burning liquid straight into our mouths. The grown-ups were usually too drunk to notice.

By late afternoon, the women and girls of Rokeby had gotten into our dancing dresses. Some of us had petticoats. My dress was yellow with a patchwork of large red and blue diamonds on the bodice. The hem had red ruffles below the knee.

A group of professional musicians—replete with a square-dance caller—set up on the front porch overlooking the circular driveway, which had to be hosed down so it wouldn't be too dusty for the dancers.

About fifty dancers lined up in two rows, partners facing each other. Everybody was clapping to the beat, hootin' and hollerin'. The end couple held hands and skipped sideways down the aisle, then back again. When they got to the end, they parted ways, one turning to the left, the other to the right. And the rows followed them, turning themselves inside out as each dancer skipped down the outside of the rows. At the bottom of the new line, the end couple formed a bridge with their hands and each couple passed under it, taking their place in the new line. The couple now at the end skipped sideways down the aisle.

I flew down the center in my dancing dress, skipping sideways between the rows with one of Dad's buddies, to hoots and calls. Our feet made scratching sounds as they kicked up the dust.

By sunset, the pig was fully roasted and ready to feast on. Some guys were carving it on a table made from a sheet of plywood resting on two sawhorses. People picked at the greasy meat with their fingers. The brown skin was crisp and crunchy. I couldn't imagine it as the same animal that, just yesterday, had squealed and panted in the junked Volvo, then hung upside down from the Chinese chestnut tree in the barnyard.

The sun went down. Little kids began passing out from exhaustion, their faces smeared with dust and tears as they wandered wide-eyed among the dancing, mingling guests, whimpering for their parents and bed.

Other stragglers had retired to the sofas in the home parlor.

It suddenly felt very late.

Mom was in the parlor, laughing and talking loudly to a small group of men, with a glass of beer in her hand. She was dressed in her white jeans, the only pants of hers that I approved of. But the white tuxedo jacket and red bandanna around her neck offset any coolness that the jeans might have bestowed. Her manner was wanton.

"And then Jacques—*hiccup*—like little Napoléon, slapped Teddy across the face!" Mom, who was only this lively when she'd had some alcohol, was laughing as she made an approximation of a slap through the air.

Dad was always the last one to leave any party, especially his own. But while he usually fell asleep midconversation and had to be woken up before taking his leave, tonight he was surprisingly animated. Overhearing Mom, he froze in midsentence. He had that rare wolflike look in his eye. "Now, if you'll excuse me . . ." He got up and approached Mom. "Ala, it's time to go to bed now."

Mom stumbled a bit as she continued to laugh hysterically, her eyes swollen slits.

One of the men reached out to help steady her, but Dad cut in, grabbing her elbow forcibly. "Ala, we're going upstairs *now*!"

I'd never seen him assert himself with Mom before.

"No!" She pulled away. "I don't have to listen to you. You're a swine!"

He started to steer her out of the parlor, holding her by the elbow all the while. She tried to pull away from his grip. "No, I won't go. Why didn't you handle Jacques like this after he whacked you?"

I followed on their heels, embarrassed not only in front of the remaining guests, but also in front of George Washington with his white horse and Pompey. At the foot of the stairs, Mom was still trying to slip out of Dad's grip like a willful toddler, but Dad was too strong. As he stood on the first step, facing backward, he held her hands and just pulled, taking a step up. Mom's feet had to follow, so she tripped on the first step, landing on her knees. Dad just kept pulling until her body straightened. Then he started to drag her.

Thump, thump, thump . . . As her body hit each dirty step, it swept it clean.

Mom was cracking up.

"Dad, stop it!" I cried.

Thump, thump, thump . . .

"Your mother's been a bad girl. She must go to her room," he stated, as if in a trance, repeating orders.

My palms were sweaty, and I felt that familiar sensation: a heavy heart, shortness of breath.

Past the crumbling faux marble and the certificate for Uncle Lewis from the New York State senate. Past the portraits of John Jay Chapman and William B. Astor.

Thump, thump, thump . . .

As she laughed, Dad said, in the chastising voice of a teacher, "It's not funny, Ala."

Thump, thump, thump . . .

"Oww . . ." I noticed that despite her laughter, Mom's face was wet.

Past the once-white banister.

As soon as Dad reached the top floor, Mom—now sobered up by panic—took off like a deer into the night, through our apartment and into her bathroom. I heard the click of the hook and eye inside.

"Mom!" I screamed, racing after her and feeling responsible. I could hear her blubbering inside.

"*Idz do dupy!*" she shouted between sobs. I knew the word *dupa* from a silly Polish rhyme Dad used to say all the time: "*Pupa, kupa, dupa. Kto zjadl moja zupe?*" This translated as "Butt, poop, ass. Who ate my soup?" Mom used to tell me not to repeat the rhyme because it had a bad word in it, but she wouldn't tell me which one it was. Now I assumed it was *dupa*.

I howled and pounded on the paneled bathroom door for Mom to let me in.

"Go to your room."

"I won't go unless you let me in first."

"Go to your room. . . . Don't worry. . . . I just need to go to the bathroom."

"I don't want to. I just want to come in." Eventually, though, I grew too tired and gave up. And I never did get to see Mom crying on the toilet seat.

I felt afraid of this cold, violent man whom I saw only very rarely. This was the same man I had seen after Grandma Claire had had Cricket put to sleep.

But I knew from that experience that once Dad had satisfied his need for revenge, everything would return to life as usual, without further comment.

CHAPTER EIGHTEEN

LIKE PROPER ARISTOCRATS

I rode around and around the circular driveway in front of the house on my pedal-brake bike. I pretended to be lost in my own child's world as I stood and pushed down on the pedals with all my ninety-one pounds.

The Catskills seemed to be crouched in a huddle with their backs to me.

The other survivors of the past summer had gone. Aunt Liz and family had returned to their urban, Parisian life, with their Ovaltine, fresh baguettes and croissants, Mercedes, school uniforms, and au pairs.

Even Giselle had returned to France with her family, allegedly for reasons relating to her husband's work. As suddenly as the hurricane had come, it was gone. But as in the aftermath of any great storm, nothing would be the way it had been before.

Mom was now hanging the underside of her long hair out to dry, over her knees, on the warm, wide stone porch steps. Dad was under a car, fiddling with an oil cap that had soaked his hands black. A few of his sidekicks were standing around observing his machinations. One-armed Roy was there, and obsequious Pete, still in his early twenties and searching for himself.

Mom—veiled by her damp hair—was pretending not to care, pretending to be asleep, or indifferent enough to fall asleep. Pretending that Dad had never uprooted her from her mountain village in the Polish Carpathians and transplanted her here in the first place. Pretending not to hear her sister-in-law as she passed with her menagerie of horses and goats in tow, dressed in her expensive, stiff black riding boots and skintight beige riding breeches. Pretending to have forgotten the storm.

If she were asleep, she wouldn't have to acknowledge anything that had happened here.

Uncle Harry's green Pinto appeared over the final rise of the carriage drive. The car sped toward the house trailed by a cloud of dust, then pulled up in front of the house. Uncle Harry slammed his car door, briefcase in hand and dressed in his usual suit and tie.

A second car followed. Two men got out.

Uncle Harry approached them with a toothy smile and gave each one a firm handshake. "Welcome to Rokeby, home to Livingstons, Astors, Chanlers, and Aldriches!"

Dad slid out from under the car he was working on. Today, the name tag on his used blue worker's shirt read ANGEL.

"Hello," he said from his prostrate position on the gravel. "You must be the fellas from *New York* magazine."

"My brother, Teddy." Uncle Harry waved a hand in Dad's direction. "We'd better get started, as I know we are only one of several Hudson River families you will be visiting today."

We had all been warned that *New York* magazine would be doing an article about the descendants of the Hudson River aristocracy who still resided on the original estates. Mom was not to be sunbathing on the front steps, Dad's friends were to be out of sight, and Dad was to be cleaned up.

I leaned my bike against the porch wall and followed Uncle Harry and the two men into the house. Mom slipped away, as she had not been invited to participate in the article.

Uncle Harry had made it very clear that only the direct descendants of General Armstrong were of interest.

"General John Armstrong, the house's original builder, can be seen in this portrait over here." Uncle Harry had stepped into his role as the family historian. "In addition to his propitious marriage into New York's aristocratic Livingston family—his marriage to Alida Livingston provided him with more than seven hundred acres of land on both sides of the Hudson River—Armstrong had an extensive political career, serving as U.S. minister to France from 1804 to 1810, and then as secretary of war in Madison's cabinet, which was his position when the War of 1812 broke out with England. He was eventually blamed for the burning of Washington and lost his job. That was the last position he held in government. When he left France, he returned with several merino sheep, a personal gift from Napoléon, for his new homestead in upstate New York. Thus the estate's original name, 'La Bergerie.' It was later renamed 'Rokeby' by his wife, Alida Livingston Armstrong, after the eponymous poem by Sir Walter Scott."

The photographer seemed enchanted by the crystal ball seated at the foot of the solid mahogany banister, whose dusty balusters dangled beneath like spiders' legs.

"Hey, Dad," I called. He was standing in the front doorway. "Tell them about how you almost killed the maid with the crystal ball!" I wanted the reporters to get a balanced picture.

"Oh, well, I doubt that's the kind of story these gentlemen are looking for." Dad's eyes shone mischievously as both our visitors hungrily insisted on hearing the story. "Okay then. One day, when I was about four, a lady employed by my grandmother to clean the house told me what a good boy I was. This was a dreadful mistake, as I felt compelled to prove her wrong. I thought I'd pretend to hit her over the head from behind with the crystal ball while she was on her hands and knees scrubbing these front stairs. I figured I could raise it over my head and begin to aim it at hers, but it was so heavy that I was unable to stop its descent at the last minute as I'd planned. The ball hit her head. I'd only managed to soften the impact. She made a horrible groaning sound. I had inadvertently knocked her unconscious! Needless to say, she never returned to work for my grandmother."

"Yes, well . . ." Uncle Harry snorted. After hinting that Dad should probably go tidy up for the photographs, Uncle Harry led the men into the reception room.

The reception room was the most formal of the front rooms. It had a set of very uncomfortable armchairs and a sofa, all with scratchy wool upholstery, unyielding stuffing, extremely straight backs, and shallow seats, arranged around a white marble mantelpiece under a mirror that extended to the ceiling. Maybe these seats had been custom-made to mold to the bodies of the early inhabitants, who perhaps had short torsos, or perhaps the ladies, in their tight nineteenth-century corsets, only sat on the edges of these seats anyway, as they sipped their tea or listened to each other read the latest Dickens installment.

One of the reporters tried to play the grossly out-of-tune upright piano. It reportedly had been part of John Jacob Astor's original instrument inventory—he had made his fortune on musical instruments as well as furs. The piano now produced a tinny twang like a starving cat. An eighteenth-century wooden flute rested on top of the dying piano. A pianoforte in similar condition occupied the other corner. In front of the French door was a white marble statue of a floating cherub, resting on a wobbly wooden stand.

"This woman looks familiar," the photographer said of a white bust balanced on the radiator.

"Julia Ward Howe—the famous nineteenth-century activist and intellectual who wrote the words to 'The Battle Hymn of the Republic'—was the great-aunt of the Astor orphans," Uncle Harry began, "and sister of 'Uncle Sam' Ward. Following Aunt Julia's example, my grandmother Margaret Chanler Aldrich had rolled up her shirtsleeves to do charitable work and fight for social and political reform—as befitted a young aristocrat of infinite wealth and energy. Grandma Margaret taught reading and writing to freedmen and -women as a volunteer at the Tuskegee Institute, under the auspices of Booker T. Washington—whom she later invited to lunch at Rokeby.

"During the Spanish-American War, Grandma Margaret joined the Red Cross as a nurse's aide and paid to have a hospital built in Puerto Rico in 1898 for soldiers wounded in the fighting in Cuba."

Uncle Harry's lecture was interrupted by a crashing sound, then Mom's voice projecting from the pantry door under the front stairs. "The icebox is empty, so you'll have to eat shit for dinner if you can't dig up any cash!" Cursing in English gave Mom a sense of power.

"Why don't you come out and speak to the reporters," we heard Dad say, knowing full well that the last thing Mom wanted was to speak to them.

"I don't give a damn who's here! Let them put in their article that we have a mansion full of marvelous antiques but nothing to eat!"

It was at times like this that I wished Mom was as skilled at pretending as I was.

Luckily, just then, Aunt Olivia bustled in, wearing Great-Grandma Margaret's black gown for the photo shoot. The dress fitted her buxom body better than it had Diana's. It featured lace frills along the waist, sleeves, and hem, and was lifted aloft by many rustling petticoats. Maggie and Diana, dressed in torn jeans and T-shirts, appeared in her wake.

"The Armstrongs' daughter, Margaret, married William Backhouse Astor Sr., the son of the first John Jacob Astor. J. J. Astor soon bought out Armstrong's shares of the estate for a reported fifty thousand dollars." Uncle Harry slipped in a last essential morsel before his wife took the stage.

"Where do you want me?" Aunt Olivia asked loudly. "We're doing this in the library, aren't we?"

Aunt Olivia knew how to carry on like a movie star, with

her face expertly made up—subdued but elegant—and her dark brown hair pulled up into a tight bun on the top of her head. Pearls dangled from her ears and encircled her neck. Although she wasn't a descendant of the Armstrongs, Aunt Olivia had ancestors among America's great railroad tycoons and bankers illustrious enough to earn her the right to participate in the article.

She passed under the overlarge portrait of Great-Grandma Margaret as a young lady, dressed in the same dress. Great-Grandma Margaret would never have allowed Aunt Olivia, a divorcee, in the big house.

For the first photo, the photographer wanted us to pose in the octagonal library. Dad and I were arranged on a chair and sofa in front of a shelf displaying a photograph of Teddy Roosevelt, as well as a bust of a Roman senator, a piece of brocade from China's Imperial Palace and a lamp with its tattered shade askew. I wished I could somehow distract the lens away from Dad's wrinkled jacket and the residue of dirt on his hands and face and under his fingernails.

Next, Dad and Uncle Harry were asked to pose for a modern-day family portrait with the younger generation on the sprawling western lawn. This was where Victorian children used to play tag while guests lounged in wicker chairs, smoking cigars, talking politics, and watching the sun set behind the distant, blue Catskill Mountains.

"Yes, we do plan to keep the property in the family." We children answered the reporters' questions.

"No, we don't see ghosts."

"No, we never met our great-grandmother."

The photographer placed me in the middle of the front row between my younger cousins, Maggie and Diana. I was holding a bouquet of purple wildflowers, slouching into myself, wearing my usual squinty smile, passive and benign. I was dressed in a plaid wool skirt, with varying shades of gray and maroon, which clashed grossly with my turquoise terry-cloth top, my pink elastic belt decorated with strawberries, and my pink hair band. Maggie's jeans were torn at the knees.

But this was just the kind of shabby detail that the author of the article wanted to stress. Here were the survivors, mismatched and impoverished, of an obsolete aristocracy, still clinging to the worn elegance of a bygone era.

"What does it feel like to be aristocrats?" the reporter asked.

"Um . . . interesting."

"This must be an amazing place to grow up!" he said, marveling.

"Yes . . ."

"Okay, I guess . . ."

"Why yes," Uncle Harry chimed in. "Rokeby is a children's paradise!"

PART V

OTHER EXILES

CHAPTER NINETEEN

HOME AND AWAY

If at first the barn fire seemed to have consumed that sum-mer's drama, it in fact made way for an entirely new kind of excitement.

With Giselle's departure, Grandma's ferocity faded, as did her sense of purpose. Robbed of those brave, violent en-counters with Giselle, she spent her lonely afternoons binge-drinking in her bathroom. By the time I got home from school, she'd be passed out on the bed.

I still preferred Grandma's house, with its TV and junk food, to the big empty house with rusty, vacant fridge.

On a cool October afternoon, I made my way from the school bus stop to Grandma's house. The bright oranges, yellows, and reds of the autumn leaves looked to me like the result of a nasty chemical experiment rather than a natural turning of the seasons.

I walked into the atrium and dropped my school bag on the salmon-colored tiles. The living room was unusually dark, its reading lamp off and its wooden blinds closed.

Assuming that Grandma Claire was sleeping something off, I made a beeline for the fridge. I took a chocolate pudding from the door shelf, turned the oven on to preheat for two squares of frozen pizza, grabbed a spoon from the dish rack, and headed for the TV at the far end of the living room.

Before I could settle into my after-school soap-opera-watching stupor, I heard the storm door slam. Mom burst in, her face full of panic and fury.

"Grandma Claire's in the ICU with alcohol poisoning. . . . The ambulance came. . . . She was throwing up blood. . . . Looked like shit! Almost died right here in the house."

Crisis made Mom manic. She spent a lot of time envisioning disasters, end-of-the-world scenarios. For her, each crisis was like a little messiah, lifting her out of her own trapped existence and into the heat of someone else's moment.

She marched over to turn off the TV. "I know you'd rather sit here and eat Claire's junk food!" Mom didn't understand that I needed to spend my afternoons here at Grandma's, that it felt more like a home than the big house. "She might not make it in the hospital," Mom added. She hadn't looked at me once with her tiny, squinting eyes, whose angry expression was accentuated by her blue eyebrows.

I tried to picture it—*vomiting blood*. Throwing up was disgusting enough, but blood? It sounded as if she'd been shot. My muscles tightened.

"What can you expect, when a person abuses their body that way?"

I doubted that Grandma Claire wanted her life to be saved.

"Let's go. You're coming with me. . . ." She grabbed the pudding container out of my hand. "What's that cooking smell? You're just like your father, with a taste for American junk food. And let the dog out, so it doesn't piss in the house." Mom was already rushing out the door with near-deadly force.

There had never been an exact moment when I suddenly understood that Grandma Claire was a drunk. I had been very lightly aware of it from watching Aunt Olivia's son Ben empty bottles of Jim Beam and Jack Daniel's down Grandma's drains, from hearing Aunt Olivia's mocking remarks about Grandma's "sucking on the bottle," and from the scent of liquor and mouthwash on Grandma's breath. My awareness, like a cake in the making, had grown gradually richer with each additional layer.

❀

Not letting the day's events interfere with their plans, Mom and Dad were going to a dinner party.

"Hi, A-lex-an-dra." My half–Puerto Rican cousin Veronique greeted me from our apartment's doorway. She was here to stay with me while my parents were out. Veronique was Cousin Chanler's twenty-year-old granddaughter. Her mother was Puerto Rican, so even though Veronique had a wide mouth and dimples, like me, her lips were much more voluptuous.

"Hi," I grumbled, uncomfortable around her positive energy. It was hard not to like her, though, with her lovely cocoa skin and soft-looking, fluffy afro tied back with a kerchief.

"Don't worry," she whispered, and winked. "Your grandmother is a tough lady. . . . Hey, can we go exploring through the house?"

"It's cold everywhere." I felt little excitement about roaming around the house. I was irritable with worry about Grandma Claire. She was so thin and fragile, so desperate to fix her family and yet so destructive. "We'll have to wear our coats. Most of the house is like a refrigerator."

We walked out into the third-floor hallway. It was lit by a single bare bulb.

"Where does this lead?" She pointed to a door.

"There's an old-fashioned elevator behind there."

"It's too bad you can't use it all the time, instead of walking up and down three flights of stairs every day."

"There are a lot of neat things in this house that we're not allowed to use. . . . Be careful!" I warned, as she opened the door to the empty elevator shaft, a dark void.

Standing on its precipice, Veronique started to pull on one of the frayed ropes dangling at the side of the doorway. At first, it was resistant and needed tugging. Then it started to slide quickly through her hands, threatening rope burn, whirring loudly as it sped along. Soon, the top of a wooden box appeared from below, then more of it, until a tiny room with a wooden bench against the back wall stood before us. It was the first time I'd ever seen the inside of the elevator.

Veronique sat beside me, still holding the rope, which had a locking mechanism so that we wouldn't go flying as soon as we sat down. Then she started tugging on the rope for us to move; we surged upward with each pull.

The elevator soon stopped. "We've hit the top," Veronique said. It was pitch-black.

"Where are we exactly?" I asked, disoriented.

"We're at the edge of the house."

"What are we doing?"

"Just sit still and feel the energy of this space. It's so quiet. I'm sure your great-grandmama never came to this part of the house. The house ghosts don't roam here."

"How do you know?"

Veronique's brother had been killed in a motorcycle accident a few years earlier, at the age of seventeen. As one of the only black kids in the local school, he had been very badly treated by his peers and had turned to drugs and risky behavior.

"Because ghosts revisit spaces they used in their lifetimes, spaces that had meaning for them. It is not logical for a ghost to explore new spaces. It is dead. It is not motivated by curiosity or desire for new experiences."

I didn't like novel experiences either.

"I guess that makes sense." In fact nothing was making sense to me.

My feet were cold in my Polish slippers, and I hadn't brought gloves; sitting on my hands on the hard wooden bench of the elevator didn't help. And I didn't want to tell her that I tended to get claustrophobic. So, with my body tense from cold and the discomfort of being in a tight, pitch-black space, I kept still.

"I think you've had a rough go of it."

"No, it's not so bad." I despised being pitied. But in truth, I was miserable. The only person I could rely on to take care of me had nearly died that day.

In fact I had more in common with Veronique than I did with most people. She understood how difficult it was to be raised by people who inherited property without money, and who themselves were raised in privilege. Both Dad and Veronique's grandfather Cousin Chanler had caught the tail end of the old glory days. They had watched as their parents and grandparents had thrown dinner parties, gone to clubs, traveled the world. So although they both got the requisite Ivy League education, they never learned any professional skills. And of course they each had too strong a sense of entitlement to do a single job day after day and take orders from others. While they believed they were independent and could do whatever they wanted, they didn't inherit the money to support that attitude. Veronique and I would have to learn how to make it like regular people. But how could we be normal when we had no normal role models?

I could, however, imagine a different life. I'd thought so much about what I called "my New York City plan" that I almost believed it was about to happen.

I would be living in the city with a rich, childless aunt who would pay for my private school education and violin lessons at Juilliard. She would have a brownstone like the one Great-Grandma Margaret used to own. I would have a small room, with heat in the winter and lots of privacy, where I could do my homework and practice my violin undisturbed.

The only problem with this plan was that all my relatives had been accounted for.

❀

When Grandma Claire finally came home after a week in the hospital, her face looked plumper and had its color back. Her eyes gleamed when she smiled. There were no more visible stress lines around her eyes and mouth, which previously had appeared so devastated by exhaustion and toxicity.

"I'm going away again for a few weeks, dear," Grandma told me. "Your uncle is sending me out to a resting place in Minnesota."

I knew Minnesota was a state, but I thought of soda.

I thought of the way Grandma Claire's facial skin looked like caked baking soda. Though she powdered it every morning with a powder puff, she missed crucial spots such as the middle of her cheek to her ear, or the tip of her nose. And I thought of the soda fountain at the old-fashioned diner and ice cream parlor in town where Grandma would take us kids. We would sit in the deep, highly varnished mahogany booth and eat grilled cheese sandwiches with ketchup. For dessert, Grandma would order a vanilla ice cream float with root beer. I never liked drinking soda because the fizz would hurt my tongue.

The news of Grandma's departure burned me in a similar way.

"Why?" I whined. Of course I understood. "Can I come?"

She had gone away before, each return beginning with hopeful expectations. I remembered one golden autumn day at Edgehill in Rhode Island, where we'd dropped her off. There had been a chill in the air. My heart sank as Dad's car pulled away from Grandma, who waved to us. Her silhouette and long shadow as she stood in the middle of the driveway are in my memory still. I continued to wave through the car's back window until her silhouette disappeared completely from view. I don't know if I saw a tear roll down her face, but I wanted to remember her lonely figure with that tear.

"No, dear. It's not for families." Her long fingers rested on her knee and tapped in a semiconscious, gentle tic—from guilt and embarrassment, I knew. "Grandma has a problem and needs to get well."

I preferred Grandma's version, which omitted the gore. Still, I couldn't get the thought of that bloody vomit out of my mind. I thought I could detect hints of its acrid smell in the house. She probably had been unable to lift her frail, six-foot frame out of the recliner and gotten some on her clothes. Had she passed out? Mom had forgotten to tell me.

"I'm going to return these to the library tomorrow. I checked them out for you before I went away." She handed me *The Life of Paganini* and "something on the Dreyfus affair, to supplement your social studies unit. . . . Of course I could renew them for another two weeks, if you don't end up devouring them by tomorrow. . . ." She pursed her lips. "Meanwhile, how 'bout a game of Scrabble?"

So much a child herself! Maybe she had made herself sick with alcohol because she, like me, hadn't really had a childhood. Maybe the alcohol took her back to it, or helped her forget it.

Dad seemed to view alcoholism as a family tradition. That evening, he reminisced about his pop's alcoholism and the many rehabs he had visited: the Institute of Living in Hartford, Connecticut; Austen Riggs in Stockbridge, Massachusetts; and Stony Lodge in Ossining.

"At Stony Lodge, they took the patients on tours of the neighboring maximum-security prison, where they would show them the electric chair. My father was a total failure in that place, and they kicked him out eventually. He always told them that he wanted to be an alcoholic. And that really cut the ground out from under his psychiatrists. Maybe Pop's drinking had something to do with the fact that my grandmother had never allowed alcohol into the house. So Pop went to the other extreme, I guess. . . .

"Pop would get so drunk! When it came time for me to graduate from high school in 1958, my mother snuck out of our New York apartment early in the morning while Pop was still sleeping something off and locked up all his clothes so he wouldn't be able to come later, as he would inevitably be too drunk to attend the graduation. So when he did wake up, he borrowed clothes from our elderly neighbor, Captain Barlow—of the British army and the Bermuda line—which were several sizes too big. Then he caught a flight up to Boston from LaGuardia, then another to Lawrence, and a cab to my boarding school. He arrived well before my mother, who was driving up. He had already gotten into the sauce on the plane, so the assistant headmaster, who was a close family friend, kindly locked him in his office until the graduation was over. You know, the odd thing is that my mother only started drinking after Pop died of it."

* * *

WHEN GRANDMA CLAIRE returned from rehab in early December, I felt hopeful that in her new sobriety, she would assert some parental authority and establish order. I even became chatty and affectionate with her again. The usual current of rage and betrayal that her drinking triggered in me had stilled.

CHAPTER TWENTY

MIGRATIONS

Boxes of books suddenly appeared in the third-floor hallway outside our apartment. There were so many of them that we could hardly pass. Old Cousin Chanler had just died. His death had left Dad with quite a funeral story, and Uncle Harry with additional archives for his Rokeby collection.

Mom had been ordered to vacate the billiard room—which had served as her bedroom for many years—to make space for these new Chapman archives. Dad and I would be moving into two of the old storage rooms, which Dad would renovate, to make space for Mom.

"Your uncle wants to keep history about that branch of the family away from strangers," Mom joked, seemingly unperturbed about being displaced.

I WENT DOWN TO Maggie and Diana's part of the house, as usual, to borrow some of their domestic stability. I found them in the middle room, reading together in their father's leather recliner.

I picked up an old Talbots catalog that was hanging out of their magazine rack. "Let's look at Talbots. Shove over." I squeezed into the recliner next to my cousins and opened the catalog.

"I call *her*," I said as I pointed at the model I wanted to be. She was, of course, the most glamorous one on the page.

"Not fair," Maggie complained. "How come you always get to be the prettiest?"

"Because I'm the oldest, and I know the most."

"I call *her,* then." Maggie was always the red ribbon behind my blue. And Diana happily accepted third place.

"And she is Anna." We assigned an older model to our absent cousin Anna, who was overseas and had no say in the matter.

Suddenly, a burst of agitation came from their kitchen. Aunt

Olivia was crying, and Uncle Harry's voice was growing louder.

Then both grown-ups rushed through the middle room, Aunt Olivia scurrying on her husband's heels.

"How could this have happened on a school trip?" my aunt sniffled, holding a napkin next to her nose.

Uncle Harry, wearing his official-business face, said nothing.

They blustered quickly down to my parents' kitchen, from which came more moaning and caterwauling. Then Uncle Harry spoke, sounding like Abraham Lincoln delivering the Gettysburg address.

". . . might be the greatest tragedy this generation has endured . . ."

We rushed to the top of the staircase that curled down to the old pantry outside my parents' kitchen so we could eavesdrop.

Behind us hung Uncle Bob's sinister painting of Death, on a canvas about ten by fifteen feet. In it, a skeleton played the flute to the marching of pilgrims up the mountains of their respective religions. In the dim light, the whole canvas looked totally black.

We tried to make out the phrases between Aunt Olivia's sobs.

". . . on a ski trip . . . neck's broken . . . may never walk again . . ."

"It's Ben," Maggie whispered, wide-eyed, violently gnawing on her thumbnail.

Diana whimpered. "Ben's going to die. . . ."

"Where is Ben now?" We could hear Dad ask.

"Mass General," Uncle Harry answered. "We're going to drive up to Boston tonight. . . . The girls can stay at Mommy's house. But just for tonight, so as not to alarm her this late, they'll have to stay with you on the third floor."

"Let's go back up to the middle room before they find us here," I suggested. We quickly reopened the Talbots catalog as if we hadn't heard a word. Except that Diana was now moaning.

The air was tinged with the excitement of a big event, as it must have been when the ambulance had come to rescue Grandma Claire from her bloody vomit.

Later that evening, Diana came up to our apartment in her pajamas, dragging along a torn mini patchwork quilt that she called her "little blankie." She didn't speak, but rather stared at all of us: Mom, Dad, Maggie, and me, sitting in a row on the green leather sofa.

My parents had never hosted Maggie and Diana before. The only time my cousins ever came up to our apartment was to play with dolls in my room. I believed that their parents didn't think our apartment was a suitable environment for their children. They were the respectable family. Thus, I would eat with Aunt Olivia and Uncle Harry and go on trips with them, but their girls never ate with us or accompanied us to the movies or the A & P.

Diana's face wrinkled up as she started to cry.

"What does the kid want now?" Dad asked nobody in particular.

The next day, Maggie and Diana settled in at Grandma Claire's house. Maggie slept on the rock-hard horsehair mattress in the peach-colored guest room. Diana slept on the daybed in the long front room. We all lived in a mood of frozen dread, not knowing how long Aunt Olivia would remain with Ben in Boston, how long Maggie and Diana would remain at Grandma Claire's, or when to expect Grandma Claire's next relapse.

I only caught snippets of information about the accident. From what I gathered, Ben had been skiing when he broke his neck. Now he was paralyzed from the waist down. What made him a quadriplegic was the fact that, as Uncle Harry explained it, he also had lost all fine motor function in his hands. None of us had been permitted to go up to see him, but I imagined him broken and stretched out in traction on a hospital bed, wrapped, mummylike, in bandages.

Shortly after Ben's accident, Aunt Olivia decided to buy Cousin Chanler's red ranch house, which would be easier for Ben to navigate in his wheelchair, and which had been subdivided from the rest of the Chapman estate after Chanler's death. They would move in when Ben returned from the hospital. In the meantime, Uncle Harry was always at work, Aunt Olivia was living in Boston with Ben, and Maggie and

Diana were staying at Grandma Claire's. This left Uncle Harry's part of the big house empty, with the feeling of a place that had been abruptly evacuated.

It was now Uncle Harry's family that had become fragmented. This would have been the perfect moment for my family to move into the spacious back of the house. What stopped us from claiming it, now that Uncle Harry and Aunt Olivia were away and planning to move to Cousin Chanler's old house for good?

Fear and indebtedness stopped us. Uncle Harry had always had more de facto rights and authority than his older brother. And while, in some ways, Dad was a rebel, he was terrified of Uncle Harry and would never openly defy him.

I knew that if I suggested moving to Mom, she'd echo Grandma Claire: "What do you have against your aunt and uncle that you want to steal their space away from them? Can't you be just a little bit grateful for all they've given you?"

We would never ask for more.

CHAPTER TWENTY-ONE

PRACTICALLY ORPHANS

By late December I was traveling like a Gypsy between the big house and Grandma Claire's at least once a day, armed with my violin and homework.

I redoubled my efforts to enforce order. With Maggie and Diana's parents away, I armed myself with an agenda for my young protégés. I had Diana, a first-year Suzuki violin student with Mrs. Gunning, on a strict practice schedule: twenty minutes per day.

"Okay. Let's take out our violins." Diana and I would open our violin cases on the cast-iron guest bed. "First we do what?"

"Tighten the bow."

"Good." Each of us would turn the small, metal screw-like part at the bottom of the "frog," or base of the bow, to tighten the horsetail hairs.

"Now what?"

Diana shrugged, her lips pursed, her eyes looking down at the floor, reminding me of my "conferences" with Aunt Olivia, but with a different cast and story line.

"We have to rosin the bow." As we rubbed the horsehair over the hard, amber-colored rosin, it squeaked and sent dust flying.

I enjoyed being the teacher.

"Okay, now let's tune."

"I can't tune myself."

"Oh, all right. I'll do it for you, but you do need to learn how." I would try to move the sticky black tuning pegs on her tiny quarter-size violin. "Okay, let's go over 'Lightly Row.' First play it for me."

She would scratch it out, plunking her fingers down mechanically. I'd roll my eyes. "This sounds bad. Try not to press the bow so much against the string. And relax your hand."

She would try again. Her half brother was half dead, her parents were away, and her overbearing older cousin was forcing her to relax her muscles for a less pressed sound. She did not relax.

Diana had kept quiet throughout this ordeal with Ben, though I imagined that she felt trapped at Grandma Claire's. Her soft skin had become blemished and oily, as had her hair, which appeared to go unwashed for weeks at a time. She had black owl eyes with swollen circles around them. Yet I envied the fact that she could show her sorrow. Unlike me, she apparently did not feel the need to maintain a façade of nonchalance.

Maggie suffered too, and again, not in silence. She would howl in her bed every night, knowing that she couldn't go home to her parents. Maggie cried easily; she would even bawl freely at sad movies. I envied her ability to cry out loud, unembarrassed. But most of all, I envied the attention Grandma Claire lavished on both of them. I understood even then that the ease with which they showed their feelings came from an innate sense of security, an earned faith that their pain wouldn't be ignored by the grown-ups in their lives.

"No. Listen as I play it. Then try to imitate. Imitation is the basis of the Suzuki method."

She tried again. More scratch. This time she missed some notes. I felt myself tense up. Lenience, encouragement, and patience were eluding me.

"No. Check the notes. You're doing it wrong." As her sound and intonation deteriorated, I swirled into a rage. "Why can't you get it right?"

I was angry at my mother for never supervising my violin practice, at my father for ruining our nuclear family, at Aunt

Olivia for mocking me, at Aunt Liz for taking my dollhouse and making me feel I had no part of Rokeby, at Uncle Harry for humiliating Dad, and at Grandma Claire for choosing the bottle over me.

Small, obedient Diana was the only person I could safely abuse with my fury, and as I yelled at her, it grew.

"No! No! No!" I hurled my own violin bow across the guest room. Diana's face wrinkled up, and she ran out of the room. When I went to retrieve my bow I saw that I had chipped the wooden tip.

"What have you done to her?" Grandma Claire rushed out of her kitchen, gritting her teeth and glaring at me like a wild dog, certain that I was bullying "this poor child," who was "practically an orphan."

If anything, sobriety, grim and real, had made Grandma Claire meaner and edgier. The clouds in her eyes had dissipated.

"Leave this house at once! I won't have you torturing your poor cousins. You know they have nowhere else to go!"

Every time Grandma Claire turned on me, I'd trust her a little less. Didn't she understand that I too was a child? And yet, what I resented most as I was getting kicked out of Grandma's house again was the way she had undermined my authority as a teacher and parental figure.

I grabbed my reversible blue/red puffy down vest. Banished, I wandered out into the blue wintry dusk, toward the ice-encrusted field north of the farm road.

They'd be sorry if I died.

I stumbled over the frosty surface of the snow. As my feet broke through the crust again and again, the powder underneath drifted into my shoes and chilled me with the cold and wet.

My eyes burned, but I could not cry.

The first time I ran away from home, I was three. I had just been punished by Mom for pushing the cork inside her treasured brown-glass bottle—rendering it unusable. She grabbed a leather belt and folded it. "You want the belt?" she asked me in Polish. Then she snapped it under my nose with a loud crack. "You smell the belt?" At which point I took off like a jackrabbit through our attic apartment. Although she chased me—over beds, under tables, around chairs—she couldn't catch me before I escaped into the cavernous halls and stairwells that lay like their own wintry landscape outside the walls of our apartment. Later, when things had settled down and while she was taking a bath, I furtively packed some things into my shiny, round aquamarine suitcase, put on my pink satin nightgown and slippers, and walked out into the snow. When I found Dad at Sonny's house, I asked him to call our friends the Johnsons to see if they would foster me for a while.

To me, Bob Johnson's family epitomized middle-class normalcy. They had a quaint yellow Victorian house near our church, with clean orderly rooms and a color TV. They ate meals together as a family. The boys—self-confident and comfortable with who they were—played baseball. Their mother had domestic skills; in fact, she had reupholstered several arm-chairs for Grandma. When I was little, I would sleep over at their house as often as I could to experience a peaceful and healthy family life without daily fights and power struggles.

But Dad, in a rare moment of cooperation with Mom, wouldn't call the Johnsons that evening. He returned me to the big house. After that, my escape from Rokeby occurred mostly in my mind, alternating between fantasies of becoming a famous violinist and of being adopted by wealthy long-lost aunts.

I now wheezed with the cold. Steam poured from my mouth and my throat burned. Each self-pitying moan hurt, yet the pain felt delicious, because it was my pain.

I did what I would usually do when I felt sad. I hummed a melody in a minor key. It sounded vaguely Gypsy, vaguely Jewish.

My boots made a crunching sound on the frozen snow, the sound of grown-ups walking on children's bones.

I was cozy in my sadness and self-pity, until my feet began to seize up from the bitter cold. I soon had no other choice but to head up to the big house.

I walked onto the flagstoned courtyard outside Aunt Olivia's and Uncle Harry's part of the house. The stones were crispy and slippery with snow and ice, and the lights were off in Aunt Olivia's kitchen. A bell jingled as I opened the door to their mudroom, which was dark except for some light from the back hall—a bulb permanently left on—that peeked through the cracks in the door frame. Coats still hung on the coatrack, and duck boots and riding boots were lined up on the floor.

My boots clumped heavily as I walked up the back staircase, where again I encountered Uncle Bob's giant painting of Death.

Suddenly a figure emerged on the stairs. It was "Bob the Ghost," the pale-faced man with the oily, long blond hair, almost as pale as his white, expressionless face. He stared at me, and I stared back—a sort of greeting in the form of the briefest acknowledgment.

CHAPTER TWENTY-TWO

INTO THE
MOUTH OF HELL

Tonight was Debbie's spring pageant. It was March, just days before my eleventh birthday. The air was frigid. Mom and I, together with about ten other volunteers, were wearing suffocating tubelike spandex slip-ons that covered us from head to toe.

We had been instructed to wriggle around inside our

spandex suits, to create the illusion of esophagus muscles digesting food as participants passed among us. We were the Mouth of Hell (or more accurately, the throat). The participants were on a mission to retrieve Persephone from her bed in the underworld, which was at the top of the hill, near the cliff on Astor Point, where we frequently picnicked in summertime.

To get in and out of the underworld here at Astor Point, the pageant participants were ferried by Chiron, played by Dad, across the river Styx—our deteriorating railroad bridge.

Now my wriggling and moaning inside the spandex felt senseless. I was uncomfortably blind, although it was so dark and moonless that I would probably not have seen any better had the spandex not been wrapped around me. Nor could the other participants see as they squeezed through us, the throat muscles.

Because Uncle Harry wanted to put an end to these pageants and would scream at Debbie every year before and after each pageant, I felt that my participation in this one was in some way a violation of the rules. I moaned and wriggled very reluctantly, afraid I would also get yelled at. It also felt as if, by re-creating Hell, we were messing with some dangerous, satanic powers.

About three weeks after the pageant took place, the dead body of one of the participants was found on the rocks under the cliff. He had been missing since the night of the pageant.

"John Martin and I went down there, after I got the call that there was a dead body washed up against our rocks," Dad shouted into the phone, like a newscaster relaying live war coverage over the din of gunfire. "We went out in a rowboat. It was cold like a son of a bitch. John's still out there keeping an eye on things."

I imagined the smashed body on the rocks at low tide, its hair matted with thick river grass. I imagined the waves of the ancient Hudson gently lapping against the body, trying to push its stubborn weight closer to shore in benign rejection: *Go back to the land.*

"Surprisingly, nobody asked any questions. The police aren't interested in doing an investigation. It was chalked up to suicide. They're used to bodies washing up on shore. There are so many jumpers, you know, off the bridge. They say he was a heavy drug user, with a history of mental illness."

We had to go to Irving Rothberg's to consult about the removal of the dead body. This was just the type of work Dad was competent at, as nothing was too disgusting for him.

In his unlit living room, his son and daughter were sitting on the furry green wall-to-wall carpet. The blue light from their color TV illuminated their faces, gaping full moons, lulled into a stupor by Richard Dawson on *Family Feud*.

"Want a TV dinner, midget?" Irving asked me.

Dad perched on a kitchen stool by the counter. "The kid loves 'em. She never knows where her next meal will come from." For Dad, our poverty was amusing, a delightful challenge. "So, as I was saying. This death is the excuse my brother's been waiting for, to put an end to artistic events once and for all and purge the property of what he deems undesirables."

"I'll bet he's scared of a lawsuit!"

"Any excuse to get wound up . . ." Dad chuckled. "Yup! That's the way it goes. Just another day at the Funny Farm . . . Anyway, Irv, keep me posted about any new corpses that come your way, and if you get anything really nice off any of 'em, don't hesitate to call!" Irving would give Dad the hand-me-downs— clothes, prosthetic limbs, even false teeth—from those he buried. While Dad wore some of these prizes, he added most of them to his collection of junk. "And I'll also be in touch if any new bodies wash up onto our shores!"

"Certainly, Mr. Theodore." Irving nodded matter-of-factly. Irving never smiled. He and Dad, with their matching overbites, looked like earnest beavers. "Now, take that kid on home, and show her how to do some laundry, if your wife won't do it for you. You've succeeded in stinkin' up my house."

With that, Irving escorted us out the back door and into his dimly lit driveway, where the truck was parked.

❧

Usually, the annual family meeting was held in August, as September was when the bulk of the property taxes were due. But in light of recent events, Uncle Harry called an emergency family meeting in early April.

I pretended to sweep the front hall while I tried to listen in on the commotion in the home parlor. I pictured Grandma Claire, Uncle Harry, Mom, Dad, and the family accountant, all seated around the gray marble mantelpiece—with their bills and receipts, reports and notebooks—talking of insurance, infidelity, leaks, accidents, missing roofs, barns burning, and bridges burned.

"Good afternoon, Miss Alexandra . . ."

I was startled from my eavesdropping by the mock British accent of Walter, who tended to creep out of nowhere, in the halls and stairwells, then fade just as quickly back into the woodwork. As he blithely descended the white stairs, I wondered what Pompey, Washington, and William B. would have thought of this, our latest boarder.

"Eavesdropping, I see. . . . Nice day at schoo-oool . . . I hope?" Walter spun his words with a murderous and calculating politeness.

"Shut up, you fairy!" The words slipped out of my mouth. I tended to treat Dad's boarders with utter disrespect, as it was easier for me to blame them for their invasions into our family life than to be angry at Dad for never setting boundaries.

Walter's beady eyes grew darker behind his thick, black-rimmed eyeglasses. His red, chapped, oversize lips tightened in anger under his oily, pockmarked nose.

"Don't you dare speak to me like that, you little bitch!"

He grabbed the broom out of my hands and swatted me across the butt with it.

Walter was different from other boarders we'd had, all of whom I'd found irritating to some degree but not truly threatening.

I rushed away from him now, before he could see that his physical attack had brought tears to my eyes.

🌿

As my father had surmised, the death of the pageant participant gave Uncle Harry the leverage he needed to ban future Rokeby pageants and the unwanted drifters that necessarily came with them. In his mind, the pageant had caused the death at Astor Point and exposed Rokeby to a possible lawsuit.

I imagined Uncle Harry's voice rumbling between the home parlor's walls.

"I always said that pageant organizer was a dangerous, manipulative woman, and that we are not insured for pageants, or whatever it is she does. And all those drifters that hang around the creamery . . . None of them have real jobs, of course. . . . Just a bunch of jackals waiting for a meal! I suppose they all use narcotics as well. That goes without saying. . . . Now that there's been a death, there'll most certainly be a lawsuit, and they'll milk us dry! I won't allow some performance artist to bring down this family!"

But these plays and festivals had breathed the only real life into the place since I could remember.

I thought of Astor orphan Uncle Bob's painting of Death playing the flute against the Tree of Life, which Uncle Harry proudly pointed out on house tours. When Uncle Bob, at the age of twenty, informed his Chanler siblings that he had decided to attempt a career as a painter, he felt he needed to defend his decision. He wrote to Great-Grandma Margaret, "To me an artist is the most peaceful man alive, peaceful and sometimes joyfully happy and nearly every day sad. . . . And their ways of thinking are not given to the world except by pictures, so the world calls them fools and empty. But they are deep-souled, high minded and brave."

While Mom and Dad believed that this icy spell in the creative life of Rokeby would soon pass—after all, this conflict was not a new one—I felt the new order, catalyzed by a death, as a permanent solution. With his edict, Uncle Harry seemed to be saying that the artists were undeserving and unwelcome because they were irresponsible and careless.

I would no longer visit the safe haven of the old Rokeby creamery.

STANDING IN THE storeroom that was soon to become my new room, I watched Dad jab at the loose plaster walls with a shovel while he talked about Cousin Chanler's funeral.

"I won't be forgetting that funeral anytime soon! The

family attorneys limoed up to the event from Manhattan, because they wanted to read the will to the children, which they tried to do, with some difficulty. Sty proved hard to find."

Sty had been hiding out at the hermitage, where he had lived ever since he'd been thrown out of his father's for falling asleep with a lit cigarette and setting the house on fire. When they finally found Sty, he was armed, so they stayed outside as they read him the will. He was reportedly screaming obscenities all the while.

I heard footsteps in the hall outside. It was Mom, coming to check on Dad's progress.

"When will the rooms be ready? I feel like I'm living in a Gypsy camp, with boxes of books coming into the billiard room and my stuff in the halls," she complained from the doorway.

"Any day now, dear. Any day . . ." Dad threw a shovelful of broken plaster onto the window chute that led to a dump truck parked on the lawn below. The dust made Mom cough and wave her hand in the air irritably.

"Oh, has your father told you the latest news yet?" she said, jeering.

"More news?" I asked.

"Entertaining stories are what your father lives for. But I doubt he'll be telling this one to everyone he runs into," Mom snorted. "*Giselle is pregnant!*" She laughed in contempt as she walked away.

"What?" Numbness climbed my legs, arms and face.

When Ben returned from the hospital later that spring, Uncle Harry and his family moved into the house that Aunt Olivia had bought from Cousin Chanler Chapman's estate.

When Maggie and Diana showed me their new rooms, I understood that they had done it. They had gotten out. Only Mom, Dad, and I remained.

CHAPTER TWENTY-THREE

A DREAM FULFILLED

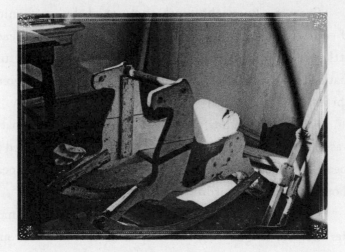

As a little girl, I used to wait by the phone, because Dad had said they would be calling any day now with a baby brother for me.

"There are plenty of unwanted kids out there. Someone is bound to call with a baby they want to give away."

And for years I'd waited, hoping not to remain an only child.

At the beginning of the summer, just as I was feeling that my own little family unit could not become more broken, Giselle reappeared. It had been almost a year since I had last seen her. As if to fill the void, she brought with her a new baby boy.

He'd finally arrived: the brother of my dreams, and the son of Dad's.

Everyone knew the truth about the baby's paternity, of course. He had a shock of straight black hair and beady little black eyes, just like Dad had in the baptism picture that had long sat in a silver frame on a desk in the octagonal library.

And yet, Grandma Claire and Mom maintained that it was best to say nothing. And since we were not allowed to speak of the facts of the case, we could definitely not speak about our feelings regarding these facts.

Giselle was physically much changed since last summer. Her shoulder-length wavy hair was disheveled in what could only be described as the lack of a hairdo, marked with a few gray streaks. The lines running along the sides of her nose and down around her mouth were more pronounced. Now that she was weighed down by a new infant, she would have more difficulty keeping up with Dad—unless, of course, she left the child with us.

When she asked to speak to me alone, I led her into the dining room. Here, whatever was said could be witnessed by the general, John Armstrong; the "Duchess," Great-Grandma

Margaret; even Aunt Liz and Grandma in their younger incarnations.

Then Giselle dropped the bomb. "I would like you to be the godmother for baby Jean. . . ."

I was Elizabethan in my icy stiffness. I had not yet even been stricken with bleeding, and she wanted to burden me with motherhood! But of course I did not say no. In my inexpressible anger, I was voiceless.

Yet I felt sorry for this infant. I wanted to rescue him from a chaotic childhood, kidnap him away from Giselle.

Giselle moved quickly. In a Napoleonic way, once she had gotten what she wanted from me—I had practically surrendered—she moved on. Later that same day, I spied her with her bundle on Grandma's doorstep, a place she'd never dared stand before. Her shadow was exaggerated, elongated in the late-afternoon sun. Her polite knock rattled against the plastic half of Grandma's screen door.

Then I saw Grandma Claire at the door, hunched over, her spindly fingers pushing against the screen door's Plexiglas. Her lips were pursed tightly. She was not quite smiling as she peered disapprovingly over the top of her black-rimmed spectacles. Yet, unable to be inhospitable within the walls of her own house, she opened the door. Giselle disappeared into the house's dimness, as if she had lasers that could burn through all obstacles, even Grandma's fury—now just a pile of smoldering ashes.

Perhaps the tragic events of this year had left Grandma

Claire too broken to fight. Or perhaps she could not deny that Giselle had won.

I slipped in through the kitchen door.

I could hear Grandma in the living room. "*Vous voulez jus de canneberge? Ou tomate?*" Even to my untrained ear, Grandma Claire sounded funny trying to speak French. "*Pas de vin,* you know, when you're nursing. . . . Is the *bébé* sleeping through the night?"

It wasn't until the day of the baptism that Giselle inadvertently introduced the baby to Mom.

Mom, Dad, and I were getting ready to go out to a lunch party at the Simmonses'.

Mom sat at our kitchen table, tapping her foot impatiently and smelling of Chanel No. 5 as she waited for Dad to show up. She had her newly hennaed hair pinned back, with some locks hanging free, like vines of morning glory around her moon face.

In a moment, we heard the echo of Dad's construction boots stomping through the front hall.

"Mrs. Simmons said one o'clock!" Mom snapped at him the minute he walked through the kitchen door. Dad was still dressed in his blue working uniform; on his shirt was a name badge reading TONY. His face and hair looked as if he'd just come out of a chimney. "Why are you getting ready only now? It's ten to one. We'll be late, as usual. Do you think I care? I should just drive myself, and let you walk. . . . But what if the brakes should fail?"

"So sorry, Ala." He turned right around, and in a moment, we could hear his boots thumping up the front stairs. In another minute, Mom's eyes squinted peevishly as another figure appeared in the dimly lit doorway.

"What does *she* want?" Mom's upper lip curled around her left canine. "We're about to go out," she shouted, as if volume would aid the quietly frantic Frenchwoman's understanding, and as if understanding would result in cooperation. We both knew she would not just shrug her shoulders indifferently and walk back to wherever she'd come from.

"Uh?" she grunted, confused. She was also squinting. "Teddy 'ere?"

"He's upstairs, getting changed. We're . . . going . . . *out*." Here was Mom, on the other side of the cultural barricade, enunciating English for the foreigner and being the mistress of the house, *belonging* to a place. The position of authority, however, was fleeting, because Giselle was gone in a flash, and we could hear her steps, as a lighter echo of Dad's, padding up the front stairs.

As soon as we heard the two sets of footsteps descend together, Mom and I headed out onto the front porch.

"Teddy!" Giselle's voice was now shrill and heated. "You said you'd come to the baptism! I *told* you it was today!" I was feeling a bit guilty, as the godmother, although it was obviously not my attendance Giselle was worrying about.

"I'm sorry." Dad's tone was now insincere, as if he were

speaking to a distant acquaintance. He walked across the driveway toward Mom's red VW Bug. "I forgot."

"Oooh . . . ," she whined, heated with agitation. "Teddy, how could you *do* this?!"

Mom followed Dad with the confidence of legitimacy in her step, her black purse dancing neatly off her hip, so light in comparison with Giselle's load. I hopped into the backseat, content that it was Giselle, and not I, who was being left out.

Giselle ran to stand between Dad and the car. "How could you be so *cold*?" she pleaded. "For baby Jean!"

"Okay, okay," he said, assenting. "I'll be there." And he brushed by her and the baby. Had he looked at the infant even once?

Mom reacted from the passenger seat. "Why are you lying? You are *not* leaving the lunch early!" Then she smacked him on the side of the head with the back of her hand, in her usual way. "Son of a *bitch*!" It was clear that her anger at Dad far outweighed any negativity she might have been feeling toward Giselle.

"But it's in fifteen min*ute*!" The usually sweet bells of Giselle's voice were clanging discordantly, as the red VW Bug pulled away with our threesome, uncharacteristically united, inside. As we left Giselle and her bundle in a cloud of dust, she trotted after us, the sack of baby now screeching, bouncing up and down.

"Teddy! Don't forget! *C'est ton fils!* Teddy! Teddy! Do you *promise*, Teddy?"

"*Sorry, sorry!*" Mom said sarcastically.

I felt sad for my father. Though he'd finally gotten the son he had dreamed of, he had not acknowledged his paternity. His long-awaited heir would remain unclaimed.

We parked at the end of a long line of cars in the Simmonses' driveway. As we crossed the wooden bridge that led to the front door, with the sound of the water rushing over the falls close by, Mom gave Dad one last jab in the side before putting on a friendly face.

"Ala!" Our hostess floated down the S-shaped stairway onto the entrance landing, with her skinny arms outstretched inside their gauzy sleeves. "It's so lovely to see you again!" Her voice was a soft wind, her phrases little gusts of airiness. She had wine on her breath.

Mom put on her demure, dimpled smile as she received Mrs. Simmons's kiss. "And Teddy . . ." Her voice was now like dry leaves crinkling under leisurely feet. "It's so wonderful to see you again. . . . Come and cool off with a drink."

Upstairs was the dining room with the balcony overlooking the sofas and grand piano below, where our student recitals took place. Various guests now stood around holding cocktails. It was quite an intellectual crowd; several professors from the local college were talking about books. I was used to such conversations and knew all these people well. I put on my intellectual façade.

"How are you, Alexandra?" I exchanged kisses and handshakes with several of the guests. "Been reading anything interesting lately?"

"I'm reading *Tender Is the Night*," I bragged. "I'm struck by Fitzgerald's beautiful writing style. I find it quite disturbing though."

"Yes, well . . . the incest, of course. Quite right."

We made our way to the dining room and Mrs. Simmons told us where to sit. "Teddy, I have you between David and Sam," said the wispy voice. Mrs. Simmons always seated Dad where his natural ability to converse would be put to best use. "And Ala next to me . . ." She was at one head, and her husband was at the far end against the glass sliding doors that exited onto a triangular balcony overlooking the stream.

Mr. Simmons was a minister. He stood, head bowed, hands resting on the back of his chair. It was strange to see him without his minister's collar. "Almighty Lord, we thank you for the food you have put before us, and for our beloved families and friends. Amen."

"Amen," we responded in unison. Then chairs scraped the floor as we all took our seats.

As the stream roared outside, the conversation circulated as lazily as the ceiling fans overhead. Beads of perspiration hung on foreheads. Wineglasses kissed lips, which were then wiped with napkins lifted from laps. Everyone smiled politely and laughed with control when a joke was told.

Dad was now telling the story of how he and his friends had gone on an archaeological expedition on the neighboring Delano estate when he was twelve.

"There was a rumor going around that slaves had been buried in a field directly south of our place, on our cousin's property, near a patch of trees in the middle of a field where there was a rock pile. My father had identified that spot as the likely location for slave burials. And so, together with the usual suspects, we decided to mount an archaeological expedition, to see if we could uncover these tombs. We thought that in the interest of science, we would not tell our cousin that we were doing this, because the scientific project was too important to be stopped by a mere landowner. So we marched over there, and we set up our camp—like any good colonial-era archaeological expedition—with a flag and tents and all that kind of stuff, and we started digging. We didn't find anything, but we dug some test holes, and we laid out our trenches in good style, figured out what we wanted to do, and then we decided we'd had enough for that day, so we camped out. Then, later on that night, my father came over and woke us up. It seemed that the superintendent over there had spotted us and turned us in to his boss. She wanted to know why we were there, and what was going on, so this caused a lot of trouble. She didn't know about the slave-bone project. So we had to withdraw; we could not proceed with the digging. And we never did find out if there were slaves buried there or not. We were hoping for jewelry, possibly gold teeth, whatever else might have been buried with them, like chains and pendants."

I suddenly noticed one of the other guests freeze as his eyes locked on something outside. I followed his gaze.

It was Giselle.

She was trying to walk quickly but gingerly down the path, so as not to trip on the stones with her bundle in her arms. She stopped about twenty feet from the glass doors behind Father Simmons and squinted into the dining room. Father Simmons's body was blocking her from most guests' view; their powers of observation had probably been dulled by the alcohol anyway.

Giselle's eyes met mine. She waved.

Facing the house's southern windows, Mrs. Simmons, whose sentences were growing increasingly drawn out, and whose breezy voice was more whistly and hissy with each glass of wine, suddenly opened her eyes wide with surprise. She waved her arm in a spastic gesticulation, as if to signal, *All eyes on me, please!*

"Excuse me, everyone." She lifted her glass as she tried to maintain composure. "I would like to make a toast. To family, and children!"

"Hear, hear! To family and children!"

"Now, who would like des*sert?* Um . . . yes. Let's see. . . . There's orange sorbet. . . . Teddy, I know you dislike fruity foods, so I also have vanilla ice cream. Oh! And fudge sauce, and whipped cream, just for you! In fact, Teddy . . . I could use your help in the kitchen."

Within minutes, I heard the front door creak open and closed. Then the figure outside retreated from view.

After the baptism, Giselle returned to her family in France with the baby. But she would pop in unexpectedly from time to time, with no regard for the lives of the people at Rokeby. Now she had the excuse she needed to come there.

"Aside from her claim, there's no indication whatsoever that the child is Teddy's."

This remained the party line.

IN SEARCH OF SELF

BLENDING IN

It was the first day of seventh grade.

Boys were playing kickball on the blacktop of the schoolyard, while girls were gathered in groups against the school building. I stood on the margin of one well-formed group of girls.

All these girls had new hairdos, feathered and layered, while only I still had the apparently unfashionable style: the long straight hair down my back, held by a barrette to each side of my middle part.

Like the runt of the litter, I was unable to squeeze into the huddle the girls formed with their backs turned outward. Although fitting in was never something that came naturally to me, I tried to laugh along at all the jokes.

They, of course, all knew that I'd always been a proud nonconformist—a dirty word in their book. It was blatantly insincere of me to be sheepishly trying to blend in now, when I'd always been at the very top of my class and self-assured about standing out.

But I'd lost that assurance. There was nowhere to go but straight into that huddle.

I WAS ATTRACTED TO a boy named Arthur, who had failed seventh grade, listened to heavy metal music, and often got into trouble at recess, for which he would be forced to stand against the wall as punishment.

But it was his suffering to which I was most drawn; his mother had recently died of cancer. Heartened by my spiritual, almost angelic role, I looked him up in the student directory and gave him a call.

"Aren't you the chick with the fiddle?" he asked.

"Well, technically it's a violin. . . ." I felt myself grow nervous about running out of things to say. "I just wanted to tell you that you can call me whenever you need someone to talk to."

"Yeah, maybe . . . Hey, what's it to you anyways?"

"I don't know."

"You're okay. But you know . . . you're not really my type. You're . . . smart. You know?" It was true that I rarely spoke to kids who weren't part of the school's honors program. "Like, you have that long straight hair, and you wear skirts. But you're kinda cute. . . ." His voice changed when he said this. Lower. "Maybe we could meet at Rocky's sometime." I panicked. Rocky's was the local roller-skating rink where junior high school kids hung out on Friday nights. I never went there.

"Yeah!" I didn't feel I could say no, since I was the one who'd called him.

"Hey, what's your phone number? Maybe I'll give you a call soon."

I suddenly felt that I was not *my* type.

Arthur agreed to "go out" with me only if I got my hair cut and styled. And he told me he was going to call me "Alex." These requests didn't seem excessive to me at the time. All he wanted me to do was step into an identity that was recognizable by contemporary standards. It was what I longed to do anyway.

I asked Christa, the girl who sat next to me at choir practice, for some advice about the best place to go to get my hair styled. Christa was popular, and her hair was, of course, *feathered*.

"Regis, at the mall," she whispered, and smiled. "Ask for Heather." She winked.

I thought about how I would look with feathers running diagonally along each side of my head. In truth, I preferred straight hair. I'd been raised to believe that worrying about one's appearance was bourgeois. And no one from my family needed to prove their worth by the way they dressed. We were above that.

At the same time, I was desperate to fit in, desperate for a friend, and even more desperate to escape myself. So I asked Grandma Claire to drop me off at the mall with some money, explaining that it was urgent for my social survival.

The salon smelled of sulfur.

"Are you sure you want to cut off all this gorgeous, thick hair?" the hairdresser asked when I told her to feather my hair.

As she clipped, I recalled how Mom used to make me a French braid, how it pinched my scalp when she pulled my hair tightly away from my face, exposing my forehead and making me look sophisticated, adult, and European.

When I returned to school the next day with my feathered hair, all blow-dried and sprayed, girls came up to me in the schoolyard, like pigeons to a crumb, smiling and telling me how "gorgeous" I looked. They touched my hair's fine feathers, crisp with hair spray.

"Do you have a comb?" One of the girls touched my butt. "Oh, no! When you have feathered hair, you just *have* to keep a comb in your back pocket! And what about mascara?"

I quickly learned to dab foundation out of its frosted glass bottle into the palm of my hand and spread it evenly over my face with my index finger: under my eyes to cover up the dark circles, over my bony cheekbones, into the crevices between my nose and cheeks, under my nose, on my forehead, between my eyebrows. And I learned to coat my eyelashes, top and bottom, with charcoal-black Maybelline mascara, draw a black line under each of my eyes, and powder my eyelids blue and silver. It was like fairy dust: instant beauty.

Grandma gave me money for black Capezio dance shoes and prewashed, skintight Guess jeans, with cropped ankles so narrow that it was a struggle to fit my feet through. I loved how tight jeans transformed me from a scrawny, insubstantial girl into someone with shape and weight.

ROCKY'S SMELLED OF polyurethane. Its atrium was dark, with multicolored disco lights scattering along the walls and floor in purples, blues, yellows, oranges, reds, as if they were little bugs also on roller skates.

Arthur and I ordered our skates.

We rolled out onto the crowded, polished floor, where the disco light scattered more color bugs. Since we were "going out," we held hands as we slipped between couples, single skaters, and groups three and four people wide. Whenever we rolled around the bend, crossing right foot over left, it was difficult not to trip.

"You wanna go in the corner?" Arthur asked. I didn't say

no, because a date at Rocky's had a protocol. This much I knew.
We pulled into a private enclave, where there was a bench to
rest on. We sat.

When Arthur put his hand on my shoulder and his face
up to mine, the image of a pregnant Giselle suddenly flitted
through my mind, and I withdrew from him in horror.

"You wanna skate some more?" I asked.

The roller skates were heavy, clunking as they hit the
floor, wheels still spinning when, with each step, they were
lifted into the air.

"Watch this." Arthur suddenly took off, speed skating.
Crouched low, bent forward, he zipped inside the ring of skaters. In half a minute, he was back next to me. I smiled indul-
gently, maternally.

Grandma picked us up at nine thirty. As she drove along
the unlit back road, with the two of us in the backseat, I felt
embarrassed. Did she know we were holding hands, despite
the near-total darkness?

GRANDMA CLAIRE REACTED to the changes in me with suspicion. She would look disapprovingly over her horn-rimmed
glasses at my tight jeans and low-cut, V-neck shirts, then purse
her lips and lower her eyes. Perhaps she forgot that she had
been the one to buy my new clothes.

"It's obvious from your clothes and cosmetics that you
are boy-crazy." This was all she needed to say. To her, boys

were depraved, and being called boy-crazy was tantamount to being called a harlot. Giselle and I were now the same in Grandma Claire's book. Just as Dad wore shirts with mismatched name tags, I had been assigned my own inaccurate label.

LOVE AND SEXUALITY had long been controversial in our family.

Uncle Lewis Chanler carried on an affair with Julie Benkard for many years, until his wife, Alice (Chamberlain), finally agreed to grant him a divorce. It was his eventual marriage to his longtime mistress that was the cause of Uncle Lewis's banishment from Rokeby, as Great-Grandma Margaret disapproved of "plural marriage."

Then there was Uncle Bob's infamous marriage to Lina (Natalina) Cavalieri, the famous Italian soprano. When he asked her to marry him, she reportedly stated, "Mr. Chanler is very nice, very kind, very good. . . . He is very rich, too, and that is a nice thing." The Chanlers were horrified by the idea of Bob's marrying Lina, a lower-class Roman who used to dance naked at a club in Paris. None of them attended the wedding on June 18, 1910.

The story appeared in the papers that three days before they married, Lina demanded that Uncle Bob sign a prenuptial agreement in the presence of a lawyer and a notary, which would transfer all his money to her to manage for him. (What he had was in trust, providing him with an annual income of thirty thousand dollars.) After their wedding, Lina decided to give him a weekly allowance of twenty dollars, at the same time that reports started to surface that Lina was seen frequently in the presence of Russian prince Paul Dolgorouki. Lina finally agreed to have the prenup nullified on condition that Uncle Bob pay her eighty thousand dollars—which he had to mortgage his New York property to get—and grant her a divorce.

When his Chanler siblings learned how Uncle Bob had been duped, they stopped speaking to him. The only one who would communicate with him was Uncle Archie, who, after his escape from Bloomingdale Insane Asylum eight years earlier, could not resist teasing his brother. In response to Uncle Bob's having once called Uncle Archie "loony," Uncle Archie wrote, in a letter to his brother, the words "Who's loony now?" This soon became the catchphrase of the decade.

Uncle Archie's love life was fraught with controversy as well. Twelve years earlier, he had fallen in love with the beautiful Amelie Rives, a writer from an old Virginia family. When it was discovered that she had based the male character in her shockingly sexy novel *The Quick or the Dead?* on Archie, the Chanlers were horrified. Archie invited none of them to the wedding. It was not long afterward that Archie's brother Wintie had Archie locked up in Bloomingdale, barring his access to his own money and property.

* * *

WHILE MY "RELATIONSHIP" with Arthur soon came to an end, my new standardized beauty improved my social standing. That is, until I came up against Colleen O'Shea.

One afternoon, I was staying after school with a new friend to watch a basketball game, when Colleen, who was a year or two older than I was, approached us in the hall. She was accompanied by two sidekicks.

"Stay away from Jimmy, you bitch!" Colleen ordered.

"Jimmy who?" I asked.

Her thick black eyeliner made her blue eyes look aglow and demonic. "You know. And if you even look at him, I'll kill you!"

I felt flattered for a moment that she saw me as a romantic rival, and I laughed lightly.

But then one of the other girls, whose mouth was too small to hold all her teeth, lumbered over to me. She pushed her face into mine. "You think I can't kick the shit outta ya?"

I attempted to move away from this girl, as I sensed that with her, confrontation of any kind would get physical.

"Hey! You listen to me when I'm talkin' to you!" She grabbed me by the back of my baggy Esprit sweater, to keep me from running. "You will learn to listen." She then began to spin me around while she stood in place and turned. I circled around her, facing outward, scurrying on my tippy-toes to avoid slipping in my Capezios. She spun faster and faster, as if we were doing a bit of ballet.

The other two were hooting. "Way to go, Joyce! Little rich bitch ain't gonna stand up to *you*!"

Then, as suddenly as she'd begun, Joyce let go. My knees hit the tiles as I fell.

"Get off your knees! You're pathetic!" And as they walked off, their cackling echoed through the empty after-school halls.

The next day, the furious threesome—who looked like a heavy metal version of Charlie's Angels, with their feathered hair flying outward like wings—blocked my passage through the school hall.

"Slut!" Colleen cried like a crazed witch as she shoved me into a locker.

"What did I do to you?" I asked, shaken. Was I a slut for having briefly dated one boy?

The bullying quickly became a routine. Just the thought of going to school would make my muscles tense, my breathing shallow, my step unbalanced, my thinking frenetic and scattered. So I tried to stay home from school as often as I could without Grandma or Mom noticing.

Clothes were now strewn all over my new bedroom floor, which still smelled of the polyurethane Dad had used when he renovated it. Like Dad, I'd stopped picking things up. What was the point of keeping my things tidy when nothing here belonged to me? I would probably be moving again, and things could be taken away at someone else's whim.

Formerly, the safety and order of school had balanced out

the dark squalor of home. But the more I stayed home from school to feel safe, the more the squalor of Rokeby overran the walls even of the private world I had previously worked so tirelessly to construct.

It was midwinter and the cast-iron radiator was always cold. Only Dad knew the secrets of our antiquated heating system, and he was usually nowhere to be found. Day after day, I tripped over mountains of detritus in my room as I'd retreat to my bed. I'd crawl under my electric blanket with a box of chocolate-covered cherries, which I would eat one after the other, cracking the hard shells with my teeth and letting the sweet liquid, slow with sugar, leak down my throat. As I'd finish the box, I'd be filled with a warmth.

This must have been how Grandpa Dickie felt in the final decade of his life, how Mom felt most days, and how Grandma Claire felt on vodka-soaked afternoons.

CHAPTER TWENTY-FIVE

INTERVIEWED

There was more and more casual mention of my going away
to school, until one day it became a serious prospect. "I
think you're good and ready to consider going away. I'll help
you make a list of boarding schools," Grandma offered.

"But what about the money, Grandma?"

"First, get accepted. Boarding schools are very selective these days. Then we'll worry about the tuition."

"Even if you *do* get into one of those fancy schools, who do you think is going to pay the tuition?" Mom asked when I told her of my plan. "And you're *not* going to get in with the grades you've been getting lately! If you *do* get in, it'll be because of family connections."

I PRACTICED FOR MY boarding school interviews in front of a full-length mirror, like a con artist working on my story. It drifted farther from the truth with each embellishment.

I wanted to convince the admissions officers that I was an intellectual.

"I know that my grades have dropped somewhat this year, but I'm sure they'll improve once I'm in a more academic environment," I told the mirror. "I'm a big reader."

I sat with my legs crossed, my elbow on my knee, and my chin resting on my knuckles, the way the genteel, literary guests at the Simmonses' lunch parties would sit when they'd speak about books they were reading or writing.

Despite my significant academic decline in the past year, I still held on to the myth of my perfection. This was why I had chosen to apply to the most elite boarding schools—Exeter, Andover, Groton. I had only one "backup school"—the Brooks School, which happened to be Dad's alma mater. I believed that the admissions officers would see through my low grades.

They would surely recognize what an accomplished person I had been, and could be again with the right support.

"I've begun to read Eastern European literature." I had recently picked up a copy of Isaac Bashevis Singer's short stories off a night table in one of the guest rooms. There was something bizarrely familiar to me about life in the Jewish shtetls of Poland as Singer represented it. The yeshiva boy—pure, aloof and studious, proper and modest—had joined the Chinese conservatory student as part of my ideal of a sheltered, disciplined life focused around the spiritual pursuits of study and prayer.

"My roots are Eastern European, you know. Oh . . . you thought I was all-American? That's only on my father's side. I grew up on an old Hudson River estate, in an eccentric family of Astor descendants who are obsessed with their heritage. My home is a mansion that was built in the early nineteenth century but is now in considerable disrepair. People tell me that it's every child's dream to grow up in such a 'paradise,' with such an interesting family, but I actually feel closer to my Polish/Russian heritage on my mother's side. I plan to return to Poland someday. . . .

"I also write. I've kept diaries for years. And I write stories and poems. . . . People often tell me I should write the story of my family's more recent history. What's so interesting about it, you ask? Although my family is directly descended from American aristocracy, my parents are rather . . . bohemian."

Reconsidering, I decided not to admit the truth about my bohemian identity to these preppy schools. I should, rather, hide the lack of middle-class order and rules, the decrepit vehicles, the inability to match clothes, the countless skinny-dipping parties, the pageants full of creative hippies dancing through the fields in costumes and carrying fruit- and vegetable-shaped pieces of cardboard, or banners that would stream through the wind. All of this was surely a liability.

I decided I should not appear too desperate for normalcy by mentioning that I had always wished I could have grown up in a three-bedroom ranch house with employed parents, siblings, cable TV, and functional cars. I still wished to convey a conservative, respectable, self-disciplined image.

My elementary school years had definitely been my best. But I couldn't very well write in my application about the story I wrote in first grade about an elephant, rhinoceros, and hippopotamus, and how impressed my teacher was by the fact I could spell these words. Nor did I tell about how I'd learned to write perfect cursive in third grade, or how I wrote as neatly and correctly as my teacher by fourth, or how I got 100s on all my long-division tests.

I did decide to mention fifth grade, my most successful year: how I won the school talent show with the Vivaldi violin concerto and a first-prize medal from the Daughters of the American Revolution for my paper on Fritz Kreisler.

"I'm an accomplished violinist. I'm working on the Bruch violin concerto and Bach's Partita Number One. . . ."

I thought guiltily about how my lessons had been going with my new violin teacher, Ms. Crowley, on to whom I had graduated after eight years with Mrs. Gunning.

Half of the time Ms. Crowley had to remind me to stop staring at my made-up face in the mirror. I had not been practicing my very challenging repertoire nearly enough. A perfectionist, she would have been the ideal violin teacher for me a few years earlier. But now, I felt oppressed by her demanding, critical style.

Watch your left hand in the mirror. Lift your elbow. Relax your shoulders.

As I would try to play better, my eyes would wander away from my fingers or notes and back to the large mirror over the fireplace. I couldn't help admiring my eyelashes, rich with mascara, and my skin, smooth and pale with foundation. I found the glamorous beauty I had achieved with makeup and a haircut reassuring, as the only achievement left to me.

"I'm not much of an athlete, but I am an advanced swimmer. . . ."

All the applications asked if I belonged to any clubs or committees. I didn't think I should discuss how, now that she was sober, Grandma Claire sometimes brought me to Alateen, a twelve-step program for children from alcoholic families.

The applications also asked me about my travels. I didn't write about the winter Uncle Harry let me come along on their family vacation to the Adirondacks, where we stayed in a heated cabin with showers and cable TV. Or Cape Cod, where

Aunt Olivia had rented a beach house one summer, and where, in Provincetown, I saw gays and lesbians publicly kissing and holding hands for the first time.

I wrote, instead, about my trips to Poland, where I believed I could be myself, where the Communists had wiped out vestigial aristocrats and old estates, and where a place like Rokeby would have been divided up into hundreds of utilitarian living units. I believed that had I grown up in Poland, I would have been both free of my heritage and free to distinguish myself exclusively through work and study.

CHAPTER TWENTY-SIX

THE ELUSIVE EDGE

In the end, the only school to accept me was the Brooks School.

It was late August and the day before Mom and Dad were to drive me to eastern Massachusetts. I slipped in through Grandma Claire's kitchen door for one final visit.

Grandma's kitchen, as always, smelled of mildew and the old brick garage. The dishes were piled up in both halves of the sink. The cast-iron skillet on the stovetop still had a layer of hard, white cooking fat. Flies were buzzing around the butter melting in its open butter dish.

Grandma Claire was sitting in her recliner in the long living room, among her musty antiques and dog-hair-covered sofas with their misshapen cushions. She was turned sideways in her recliner to view the fuzzy black-and-white lights of the TV on the shelf behind her, with an open book on her lap. The dusty wooden blinds, which would clatter when moved, were closed against the late-summer sun, their flat, horizontal pieces directed at a slant to send the sunlight to the floor.

"Hi, dear . . ." Grandma's speech was slow. She was thin and diminished. Still, we played the friendly "hi" game.

Maggie and Diana had moved away to the ranch house on the neighboring estate, and now I too was leaving Grandma, alone with her photo place mats and granddaughters' fading art projects still hanging on her walls. I could see that she had begun to adjust to this new phase of her life, one of solitude and sober old age. She seemed more stuck in her chair, more attached to this dark room, with its antiques, photographs, fireplace, disarray of magazines and papers on the dining table, lumpy sofas, and ragged rugs. Grandma Claire's head was now leaning against her chair, and her eyes were drooping. She had no choice but to loosen her grip.

A daddy longlegs moved slowly, barely perceptibly, in the corner.

I could eat supper here, but Grandma hadn't prepared for

a guest. I could set up an organized spot, an island in the sea of papers on the dining table, with a place mat, plate, and silverware. But, feeling guilt and sadness about abandoning Grandma, I had no appetite. I looked at one of her photo place mats that held pictures of relatives both close and distant, made meaningless to me by excessive familiarity. Aunt Janet in her wedding dress, Uncle Steve waving from behind a ship's helm, Grandma Claire as a wildly curly-haired infant on her father's lap, all her granddaughters in our square-dancing dresses under the maple on the western lawn, and so many more that had been ingrained in my memory from long study during dull dinner table conversation.

"Now, don't forget that it is thanks to your aunt Liz that you are going to this school. She contributed a substantial sum toward your tuition. I want you to remember that the next time you feel the urge to complain about her taking things from you."

Why hadn't anyone told me that Aunt Liz was involved? This wasn't the way it was supposed to be. Going away to boarding school was supposed to be a liberation from my family. It was supposed to be Grandma Claire who would rescue me. Now I would have an unpayable debt—like the one Dad owed to Uncle Harry.

As Grandma Claire walked me out after my visit, the late-afternoon sun elongated her shadow on the front stoop. She stood in the doorway, hunched over, her spindly fingers pushing against the screen door's Plexiglas.

"Well . . . Bye, dear. Write often. You'll be home before you know it. Thanksgiving is just around the corner." Grandma was already an expert at sending children off to boarding school.

When I tried to hug her—which I didn't think I'd ever done before—her shriveled body recoiled from the physical contact. She patted my shoulder with her gnarled hand instead. "I'll come visit you soon." Her kiss barely touched my cheek. She waved as I walked away, then called Bianca in off the stoop and let the door slam shut.

❦

Other than Grandma Claire, there seemed to be no one around to send me off—no friends, relatives, balloons, cakes. The inanimate objects in the big house had always been more present for me than the human beings.

I decided to climb the big house's tower and bid good-bye to Rokeby from a great height.

To reach the fifth floor, I walked through the library and along the creaky back staircase with its collapsing plaster walls and dangerously low banister. The room just beneath the tower was the old schoolroom, where the Astor orphans used to have their lessons with a hired tutor—until they were old enough to be sent away to boarding schools. In the middle of this old schoolroom was a metal spiral staircase with wide wooden steps. I ascended, crunching dead wasps underfoot.

Rokeby stretched all the way to the Hudson River and beyond—we even owned some of the land under the river. To the north was Cousin Chanler's estate, and to the south, the estate that had once belonged to the Delanos.

Around the tower, some crows were now coasting on the wind like kites, flipping sideways, then flapping to straighten themselves again, seeking their balance. I felt the warm wind hitting my face, blowing through me and softening me momentarily. It was like swimming underwater. I wished it could blow through the house as well and gently undo the loneliness that ruled there.

From here, I could see the parts of Rokeby I loved and knew so well—the forest paths and streams; the niches in the brush where I used to hide and chase wild rabbits; the giant, craterlike puddle in the barnyard in whose oozing mud my cousins and I used to bathe after rainstorms. I loved those days—before I'd grown stern and angry, before I'd turned my back on the squalor at Rokeby as an enemy against which I felt compelled to build a fortress of order, hygiene, and self-discipline. I loved those days when my cousins and I used to run around the property all summer long, unsupervised, shirtless, barefoot, wild—little orphans all.

The Rokeby trees, favorites of generations, were dancing in the wind. The languorous, sprawling ginkgo, with its many tiers in which we kids would lounge and read for hours, waved its long branches. The giant white pine, which had provided shade for countless picnics and croquet matches, tried to bend its trunk slightly in a courtly bow. The double white cedar, all alone in the field between the house and the river, swayed in sad good-bye.

THE NEXT MORNING, grasshoppers were buzzing in the surrounding fields. The cool Catskills were silent as I struggled to stuff my bulky trunk through the back door of Grandma's Plymouth.

By eleven, we were ready to depart. I was dressed conservatively, in a kilt and blazer.

I'd tried desperately to make Mom and Dad look fashionable. Dad was wearing a suit, which I'd insisted on preapproving. I'd had to work hard to cover up his dirty white shirt collar with his jacket collar. Mom had recently cut her hair short. Today, she was sporting punky sunglasses.

Initially, Mom hadn't wanted to come along, but she agreed after Grandma Claire had explained to her that the presence of two parents would make a better impression. And Grandma wanted me to start off on the right foot.

Giselle and the baby hadn't affected our ability to act like a unified family when circumstances required it.

As we rolled along the carriage drive, I was reminded of a recurring dream that featured this driveway.

In this dream, I would walk toward the front gate. My right leg was shorter than my left, so every time I'd step on it, I'd dip low. Then I had to drag my longer leg along the ground, position it, and heave my body up again. *High, low. High, low.* I was privileged yet impoverished, cultured yet squalid, past yet present. My family was united by a common heritage and property, yet torn apart by alcoholism, competition, and infidelity. I had so many caregivers yet so much neglect. I strove to find the edges that defined me, but the lines remained blurry.

And in the dream, the driveway had changed. Instead of the winding gravel road with mild inclines, it was a series of endless, rolling hills. Each time I'd climb another hill in my crippled state, a new one would rise up before me like a wave. In this way, the edge of Rokeby kept eluding me, and I never did reach the public road.

❁

Dad was exuberant about revisiting the institution he considered to be a great "molder of men." Mom, on the other hand, was in her usual foul mood and kept interrupting Dad as he tried to recall his own boarding school experience.

"When my mother decided to send me off to my first boarding school, I was ten. I was simply informed one day that I would be going, and was dropped off the next."

"You couldn't have taken a bath before going on this trip?" Mom complained as she rolled down her window.

"So sorry, dear . . . ," Dad said in his appeasing voice. "I thought I had taken a bath this morning."

"A bath without soap doesn't count."

"Of course, dear. I hadn't realized. . . . Speaking of bathing, at this same boarding school—where things were rather bleak and didn't really get better with time—the students had obligatory showers and haircuts. When one master complained that my hair was unclean, I got even with him by rubbing waste oil into my hair so as to make it waterproof. It was black and smelled terrible, and wouldn't wash out!"

"Ugh!" Mom moaned. "Why *ever* did I agree to come along? I think this whole business is disgusting. Someone could come up with twelve thousand dollars for boarding school tuition, while all these years we've gone hungry! And why? Why, I ask you? Because boarding school is more necessary than food in this family?"

Dad continued, unfazed. "There was one master, Mr. Jones. . . . He was particularly cruel to the boys he found pathetic or unattractive. He'd hit them with books. There was one boy named Gooding whom this master didn't like, so he used to say, 'Gooding, you're nothing more than a great suet pudding!' "

"Suet pudding?" Mom made a contemptuous grimace. "Was that something your ancestors used to eat? Ridiculous!"

"Mr. Jones would insult the boys until they cried. If they complained, he'd write out demerits for insolence or insubordination. One time, I decided to read Lord Charnwood's biography of Lincoln and write a report on it, because my pop had been reading it. When Pop complained to Mr. Jones that he thought he'd given me too high a grade on the report, Mr. Jones replied, 'I wasn't giving *him* the grade. I was giving *Lord Charnwood* the grade!'"

Mom addressed me. "And you think *we're* bad parents? You'll appreciate us after spending some time in a school where the teachers will hit you and call you all sorts of names!"

"They're not allowed to hit kids anymore, Mom," I said with a sigh.

"Oh yeah? You'll see."

It was clear to me that Mom didn't want me to go away and she thought she could scare me out of it. I couldn't understand why she cared, as she spent so much of her time alone and so little of it with me.

To alleviate my own guilt about abandoning Mom to the devastating loneliness of Rokeby, I rationalized that I needed a fine education in order to make the fortune necessary to rescue her one day.

"The school nurse was also cruel, and a drinker," Dad continued. "She was a large, imposing, mean woman, who gave injections and enemas with relish. She and her husband both drank, and after eight P.M., they wouldn't come out of their apartment, where they lived as the dorm parents of a certain dorm. They would also lock the bathroom at night, so it was necessary to pee out the window, or into a boot to be taken out in the morning. In the winter it was a problem because the dining room was right under this dorm and the pee would run down the windows and freeze yellow."

"So that's where you learned your nasty habit of peeing out your window onto the gutter!"

MOM WAS AWED into sedation by the abundance of limousines that were delivering students' trunks. "The rich kids fly up, mostly from Manhattan, and arrive separately from their luggage," Dad explained.

"Why can't I be rich like that?" Mom asked.

Not wishing to keep Mom and Dad here a moment longer, I remained distant as they got back into Grandma Claire's car. It was essential not to display to the other students anything from our life together. Any incriminating language or behavior would taint this clean slate I'd been given.

"Not everyone is so lucky, I guess," Dad—ever viewing himself as the luckiest man alive—said sarcastically.

I felt very lucky that day.

The Brooks School looked like a country club. Its sprawling lawns, its air of affluence and privilege, filled me with a sense of unlimited possibilities. Here, I would be free from the shame and chaos of Rokeby, while I could use the more glorious aspects of my heritage to my advantage. After all, as an Aldrich, I belonged here, among other students with famous names, most of whom hailed from Manhattan's Upper

East Side, where they had attended day schools like Chapin and Buckley.

As my parents pulled away, I envisioned the self-assured and intellectual person I would become here. Of course Rokeby, like a lonely orphan, would inevitably call me back, but for the moment I was free.

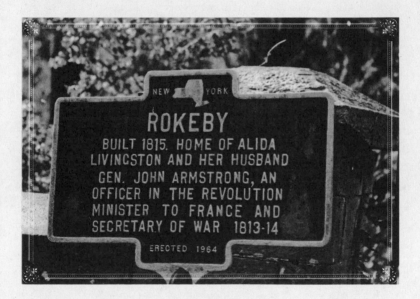

ACKNOWLEDGMENTS

I want to thank Jeanne Fleming for helping me make the necessary contacts to get this book published, and for spending endless hours on the its photos. And Harlan Matthews for supporting me through the whole process. Enormous gratitude to all my editors: Ann Patty, whose edits helped make the manuscript fit to be sold; Daphne Abeel, for believing in the book's potential; Hilary Redmon of Ecco, for her extraordinary vision. My agent, Joy Harris, for being a terrific advocate. My father, for sharing his great stories, and my mother, for sharing her collection of photographs. And last but not least, my son, Shlomo, for his patience during the writing of this memoir.